NEUROPOLIS

A Brain Science Survival Guide

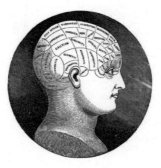

Robert Newman

WILLIAM
COLLINS

FOR YANA AND BILLY

William Collins
An imprint of HarperCollins*Publishers*
1 London Bridge Street
London SE1 9GF

WilliamCollinsBooks.com

First published in the United Kingdom by William Collins in 2017

22 21 20 19 18 17
11 10 9 8 7 6 5 4 3 2 1
Text © Robert Newman 2017

HB ISBN 978-0-00-822865-1

Edited, designed and typeset in Adobe Garamond Pro and Caslon by Tom Cabot/ketchup
Printed and bound in Great Britain by Clays Ltd, St Ives plc.

MIX
Paper from
responsible sources
FSC
www.fsc.org
FSC™ C007454

LIST OF CONTENTS

INTRODUCTION

To read the current crop of brain science books and articles is to discover that we live in 'a colourless, odourless, tasteless, silent world',[*] where 'smiling evolved from an aborted snarl,'[†] where 'Japanese people struggle to tell the difference between fear and surprise,'[‡] and where 'there is nothing special about human brains that a sufficiently complex computer couldn't do just as well'.[§] So much so, in fact, that some suggest we will soon be able to 'upload consciousness, escaping the biological wetware from which we have arisen'.[¶]

This sort of talk slanders and libels us but it is also very funny with its runaway extrapolations that leave science far behind. In fact this book grew out of a stand-up comedy show called *The Brain Show*, which toured for a hundred gigs, and then developed into a BBC Radio 4 comedy series.

---

[*] David Eagleman, *The Brain: The Story of You*, 2015.
[†] V. S. Ramachandran, *Phantoms in the Brain*, 1998.
[‡] Dick Swaab, *We Are Our Brains*, 2014.
[§] Brian Cox, interview by Hannah Devlin, *The Times*, 6 September 2014.
[¶] David Eagleman, *The Brain: The Story of You*, 2015.

My argument in this book is that brainless interpretations of brain science are doing our heads in more than we know by giving us a dehumanising and pessimistic picture of ourselves. This picture, I argue, derives not from science at all but from philosophical stowaways. Indeed if we look at what the latest neuroscience actually tells us, then a very different picture emerges.

'But who are you to talk about any of this?' was one interviewer's opening question to me on live national radio. I opened and closed my mouth like a roach on a riverbank. Minutes passed. I just didn't know what to say. I never did come up with a reply. Who am I indeed to trespass on the brain scientist's bailiwick?

In his 1940 lecture series *Dynamics of Psychology*, however, German psychologist Wolfgang Köhler praised 'trespassing as a scientific technique', on the grounds that what is merely special data in one field may turn out to have much broader significance in another. Now this doesn't mean the trespasser sees the big picture in a way that eludes everyone else. Trespassing can be helpful by accidentally treading spores from one field into another, where they unexpectedly start fizzing and wriggling into life. Or the trespasser might find fertilizer sacks full of rubble and rusty cogs blocking the entrances to badger setts. Certainly one of the great joys of researching this book has been to disinter fascinating brain science buried under all the reductive bluster.

And then there's the fact that brain science appears to have arrogated to itself all understanding of human behaviour anyway, which makes it kind of hard to move a muscle without trespassing. In fact, since the brain science fiefdom now includes life, the universe and everything, the question is who is the real trespasser here? In the words of the great comedian Michael Redmond:

> People are always saying to me, 'What are you doing in my back garden?' To which I reply: 'What are you doing in my house?'

Let's go and climb the back steps and see what they are doing in our house.

1. VOXEL & I

From the get-go, it is important to remind ourselves that brain-imaging does not actually film your brain in action. There is no live action footage of thoughts or feelings. No one will ever be able to read your mind – except your mum. Brains do not light up during functional magnetic resonance imaging (fMRI) and electroencephalography (EEG). Strictly speaking fMRI and EEG are not techniques of brain imaging but of blood imaging, since they track blood flows to different brain regions on the working hypothesis that active neurons devour more oxygen, and blood is the brain's oxygen delivery service.

On 17 May 2016 the *Proceedings of the National Academy of Sciences of the USA* published the first comprehensive review of 25 years of fMRI data[*]. The conclusions were damning:

> In theory, we should find 5 per cent false positives … but instead we found that the most common software packages for fMRI analysis … can result in false-positive rates of up to

---

[*] Anders Eklund, Thomas E. Nichols & Hans Knutsson, 'Cluster failure: Why fMRI inferences for spatial extent have inflated false-positive rates', *PNAS*, 2016.

70 per cent. These results question the validity of some 40,000 fMRI studies and may have a large impact on the interpretation of neuroimaging results.

In 2009, the journal *Perspectives on Psychological Science* published 'Puzzlingly High Correlations in fMRI Studies of Emotion, Personality and Social Cognition'. (Original title: 'Voodoo Correlations in Social Neuroscience').[*] Two of the paper's authors, Ed Vul and Harold Pashler first became suspicious when they heard a conference speaker claim he could predict from brain-images how quickly someone would walk out of a room two hours later. This had to be voodoo.

Imagine a scenario in which roaring floodwaters smash the office windows. Your colleagues flee, but then turn back to see you stranded in the rising water.

> 'Save yourself!' they cry. 'Run for you life!'
> 'You go on ahead', you holler back. 'I'm not gonna make it. I decided a couple of hours back on an airy saunter through the doorway.'
> 'Well, sashay for your life! Mince like you've never minced before!'

Vul *et al.* set about re-examining the data. They surveyed the authors of 55 published fMRI papers and found that half acknowledged using a strategy that cherry-picked only those voxels exceeding chosen thresholds. These cherry-picked voxels were then averaged out as if they were the average of all voxels,

---

[*] Ed Vul *et al.*, *Perspectives on Psychological Science*, 2009.

not just the ones that fit the hypothesis they were supposed to prove. This strategy, says Ed Vul, 'inflates correlations while yielding reassuring-looking scattergrams.'

Voxels are the organisation of statistical correlations into cuboid 3D pixels. Each cube represents a selective sample of billions of brain cells. They provide a computer-generated image of what brain activity would look like if cherry-picked statistics matched raw data. Together the cubes build a Minecraft map of the mind.

To form each cuboid voxel, you collate all the neuronal clusters that have a blood oxygen level of x at split second 0.0000001 with all the ones that have a value $\dot{x}$ at split-second 0.0000007. Junk all the non-x brain cell activity going on between 0.0000002 and 0.0000006. (Call it 'noise'). Now amalgamate your cherry-picked voxel with other voxels, (themselves boxes of cherry-picked data), and there you have your fMRI picture showing which region of the brain spontaneously 'lights up' when we are thinking about love or loss or buying a house. There you have the murky world of the technicolour voxel.

Ed Vul *et al.* call this strategy 'non-independent analysis'. To illustrate how this strategy inflates correlations they used it to show how daily share values on the New York Stock Exchange could be accurately 'predicted' by the recorded fluctuations in temperature at a weather station on Adak Island in Alaska. Here's how it works. Non-independent analysis simply skims the strongest correlations between each of the 3,315 stocks being offered on Wall Street, and finds a handful whose value appears to strongly correlate with the previous day's temperature drops on the windswept Alaskan tundra.

'For $50, we will provide the list of stocks to any interested reader,' wrote Ed Vul. 'That way, you can buy the stock every morning when the weather station posts a drop in temperature, and sell when the temperature goes up.'

Not long after came the banking crash. It turned out that Wall Street had been using some 'non-independent analysis' of its own. In the same way that fMRI false positives are boxed up into cuboid voxels, Wall Street was boxing bad debt into mortgage bonds and 'collateralised debt obligations', or CDOs – better known as the wobbly stack of Jenga blocks Ryan Gosling eloquently demolishes in *The Big Short*.

And yet ponzi voxels were to rescue ponzi bonds. Newsrooms used brain-imaging data to explain the banking crash. Neuroscience helped shift the blame from banks to brains, and from rich to poor. It turned out that the system of short-term greed that caused the banking crash was the limbic system. Neuroeconomists popped up on the nightly news to explain that sub-prime mortgages were entered into by people who let their limbic system's urge for instant gratification triumph over the prudence of their prefrontal lobes. Those who lack the mental strength to resist the limbic system's short-term greed, it turned out, would always make bad property investments. Grotesquely, ponzi voxels rescued ponzi bonds by shifting the blame onto the feckless poor.

A huge jar of sweets in a sweetshop window

To want to understand human behaviour is a human need, but frustratingly the answers are always complex and incomplete. There is no royal road to the truth, just a multiplicity of weakly-acting causal pathways. And so when we are shown a new technology that appears to answer our deepest questions, it is only human for us to want to fill our boots. EEG and fMRI are what we have been looking for all along: shiny machines that produce simple answers to complex questions. Better yet, these answers come in the form of vivid arrangements of 3D voxels, like a huge jar of sweets in a sweet shop window.

In the rush for a quick-fix answer to a complex problem did any neuroeconomist or *Newsnight* presenter ever think to blame their own limbic system for overpowering the prefrontal cortex? Why wasn't the way they themselves snatched at simplistic answers symptomatic of short-term neural reward circuitry?

One of several experiments to which neuroeconomists alluded to in the wake of the banking crash was an investigation into 'neural reward circuitry', which measured blood oxygen levels in different brain areas when people were offered five dollars now, and when they were offered forty dollars six weeks from now. The instant five-dollar cash offer represents the sub-prime mortgage. But this is the economics of the Wendy house. 'Only a behavioural economist,' says philosopher and neuroscientist Raymond Tallis,

would regard responses to a simple imaginary choice [$5 now or $40 later] as an adequate model for the complex business of securing a mortgage. Even the most   foolish and 'impulsive' mortgage decision requires an enormous amount of future planning, persistence, clerical activity, to-ing and fro-ing, and a clear determination to sustain you through the million little steps it involves. I would love to meet the limbic system that could drive all that.*

---

* Raymond Tallis, *Aping Mankind*, 2014.

* * *

I am keen to draw a sharp distinction between MRI's medical applications and its use in waffle about the neural basis of poor investment decisions and the like.

Brain-imaging helps oncologists track the success of different treatments in halting the spread of brain tumours. MRI can show the rate at which dementia is progressing. It can be used to assess the extent of damage caused by a stroke and to predict the likely recovery of brain and body function. I would probably not be able to walk but for MRI. Thanks to the magnetic resonance imaging machine at London's Royal Free Hospital, surgeons could tell at a glance that they needed to perform an emergency discectomy and laminectomy on my spine. The Registrar told me there was a two per cent chance that I would emerge from surgery doubly incontinent, in a wheelchair and in unbearable agony for the rest of my life. Sign here. But thanks to the skill and expertise of the surgeons, and thanks to magnetic resonance imaging showing them exactly where to go and what to do when they got there, I am back on my feet.

None of the medical applications just mentioned involve voxels, those 3D pixels made from crunched numbers. The voxel is a monument to the confusion of mythology with science. The wonderful medical uses of MRI lend credibility to all the mythologising.

It's only twenty-five years since brain-imaging got going. You might think the novelty of brain-imaging would make us less prone to mythologise the brain, but in fact it makes us more so. We have been mythologising the heart long enough to know when we are doing it. We don't confuse love hearts for real ones. With brains it is different.

We mythologise the brain by stashing philosophical stowaways in the uncomplaining hideouts of the *nucleus accumbens* and *ventromedial striatum*. These philosophical stowaways include for example, the revival of out and out predestinarianism that you find in *We Are Our* Brains, the 2015 international bestseller by renowned Dutch neuroscience researcher Dick Swaab:

> our levels of aggression and stress are set before birth for the rest of our lives.

I don't know about you but I was feeling pretty laid-back until I read that. Let us examine some of the other way out claims made in *We Are Our Brains.*

2. ON RAFTS ACROSS
THE SEA OF OKHOTSK

We Are Our Brains is written by a man with a grudge against humanity on account of being called Dick Swaab. Dick Swaab argues that people from Japan and Papua New Guinea struggle to tell the difference between fear and surprise:

> Japanese and New Guineans find it difficult to distinguish between a face expressing fear and a face expressing surprise.

A thought experiment seems in order.

Let's say, Yoko Ono is having her annual business meeting with Paul McCartney to settle the Beatles estate. If, during the course of that meeting, she finds herself struggling to decipher what exactly Paul McCartney's facial expression might possibly mean, then she is no different from the rest of us, who have shared her perplexity ever since that day about a dozen years ago, when McCartney went into his plastic surgeon's and said: 'I'm tired of expressing lots of different emotions, can you give me just a rictus of mild surprise and vague curiosity?'

'Sure,' replied the surgeon. 'Do you want a hint of disingenuousness with that?'

'I don't think that'll be necessary do you?'

Paul McCartney seems to get by pretty well with just the one emotion on his face. In live performance, however, he concedes that the rictus of mild surprise and vague curiosity has changed the emotional register of the songs. As he told one interviewer:

> If you take a song like 'Eleanor Rigby', when we did it with Beatles it was always very much a song about pity and compassion. Now, when I perform 'Eleanor Rigby' live, it's much more a song about mild surprise and vague curiosity. Sort of, '*Ooh, I wonder where all those lonely people came from all of a sudden?*'

The argument that the Japanese cannot tell fear from surprise contradicts one of the central tenets of human evolutionary biology. 'I have endeavoured to show in considerable detail,' wrote Darwin in *The Expression of the Emotions In Man and Animals*:

> that all the chief expressions exhibited by man are the same throughout the world. This fact is interesting, as it affords a new argument in favour of the several races being descended from a single parent-stock, which must have been almost completely human in structure, and to a large extent in mind, before the period at which the races diverged from each other.

Everything significant about our species was already well in place 35,000 years ago when Paleolithic sailors rafted across the Sea of Okhotsk to become the first humans to make landfall on the Japanese archipelago. If Dick Swaab is going to take a sledgehammer to the Darwinian principle that all people every-where express emotions in pretty much the same way, then we might reasonably expect him to provide some evidence. I mean, that's a mainstay of human biology. But Dick Swaab produces no evidence to support his claim. None. In fact, the atrocious allegation that the entire Japanese nation suffers from a sort of autism is made in a book which offers *no sources or footnotes at all.*

The New Guineans are also supposed to be unable to do what is child's play for Africans, Europeans and continental Asians, and tell fear from surprise. And Dick Swaab has worked out why. It's because: 'linguistic and cultural environments … deter-mine … how facial expressions are interpreted'.

Over 800 different languages are spoken in Papua New Guinea and West Papua, and they not even from the same language families. The Ternate spoken in West Papua is from a different language family to the Austronesian and Papuan languages spoken in Port Moresby. Nowhere else on earth exhib-its such linguistic diversity. Nowhere else on earth, therefore, is it less likely that a common language could create a shared inabil-ity to read facial expressions. Dick Swaab literally could not have chosen a worse example from the face of the earth than 'New Guineans' to support his argument. But Dick Swaab is on a roll. Don't stop him now:

> When surveying a scene, Chinese individuals, unlike Amer-icans, don't focus on a single object at a time but look at it in relation to its surroundings.

Last time I looked, the United States of America was a new-ish political state created from every race and nation on earth. According to the US Census Bureau, more than a fifth of the population, over sixty million people, speak a language other than English in the home. Americans are not a biological entity. They are not a linguistic one either. There is no specifically American way of seeing, just as there is no Chinese way of seeing. The Chinese people are not a Terracotta Army all facing one way, all seeing everything holistically the whole time. When NASA astronaut Mae Jemison, of mixed East-Asian and African-American descent, looked out of the Space Shuttle *Endeavour*'s window did she see the big picture or the small?

In her memoir *Find Where The Wind Goes*, Mae Jemison wrote that 'science provides an understanding of a universal experience'.[*] What is so terribly damaging about Dick Swaab's parascience is precisely its denial of the universality of human experience.

---

[*]   Mae Jemison, *Find Where The Wind Goes: Moments From My Life*, 2001.

3. HUMOURING THE CHAMELEON

A defining characteristic of many brain science books is a macho and rather sadistic nihilism. In *The Brain: The Story Of You*, Professor David Eagleman lords it over us puny mortals:

> What if I told you that the world around you is an illusion, an elaborate show put on by your brain? … If you could see reality as it really is you would be shocked by its colourless, odourless, tasteless silence.

I actually had to do a book festival debate with this guy. He was representing science, progress and the light of reason, and I was there for balance. At one point he turned to me and said : 'What if I told you that in the real world sound doesn't actually exist?' To which I replied: 'Pardon?'

Now I know many of you are smitten by counter-intuitive ideas such as a colourless, soundless, odourless world, but I urge you to reflect that a colourless world is incompatible with natural selection, because you have left yourself no way to account for the survival of the chameleon – unless you also believe in the existence of extremely patronising birds of prey. How kind of nature to select for hawks and shrikes prepared to humour the chameleons' pathetic attempt at camouflage! Yet how wasteful of nature to select for chameleons with subcutaneous photonic crystals! All the energy that the male adult panther chameleon (*Furcifer pardalis*) spends on lugging these photonic crystals around is energy wasted, because in a colourless world they confer no selective advantage whatsoever.

If Eagleman is right and it is indeed a colourless world, then the whole science of co-evolution must be wrong, everyone from Darwin first observing the selective advantage of a scarlet throat pouch for the Great Galapagos Frigate Bird, down to and including a 2016 *Nature* article headed 'Lizards Tailor Tails For Local Predators'. The article was about how, in three different areas of Japan, the same species of skink lizard, *Plestiodon latiscutatus*, has evolved a different coloured tail in response to different predators: for weasels a blue tail, for snakes an ultraviolet tail, for birds a brown tail. There was an interview with the project's chief scientist Dr Takeo Kuriyama, who said:

> When I first told my colleagues that I'd discovered a link between the blue-tailed lizard and the weasel they were frightened for their very lives, or pleasantly surprised. Hard to tell. One or the other.

Brightly coloured *Plestiodon latiscutatus* tails evolved to attract snakes and weasels to the one disposable part of its body. The lizard can shed its tail with the snake's fangs still in it, scuttle away and grow a new one. Ditto weasel. But birds have such sharp sight they are less likely to fall for this misdirection trick. If they see a lizard's tail they'll see its head and belly too. Against birds, the lizard's only hope is not to be seen in the first place, and so the brown tail has evolved as simple, old-fashioned camouflage. It's the least likely colour to be seen among the sticks and twigs littering the forest floor.

Imagine an organism that actually had to live in Eagleman's nuclear winter wonderland. In a colourless, odourless world, how could the orchid attract the bee it needs for pollination? You might argue that all that is needed for pollination to occur is for just one bee to land on one orchid however accidentally. Maybe it just lands on the orchid for a rest. Once the bee has found the orchid's pollen, it would then return to its hive, perform its waggle dance to inform the other bees where the pollen is at, and the other bees will follow it back to the orchid. To this argument I say: *Prasophyllum fimbria.*

The orchid *Prasophyllum fimbria* offers *nothing but colour.* From a distance *Prasophyllum* displays what looks like a pollen-spattered anther. But there is no pollen, no food reward for the bee at all. The blotchy yellow splodges are a 2D *trompe l'oeil.* The bee gropes around for ages, trying to find something that lives up to the tasty promise of the picture that drew him in, like someone eating in Harry Ramsden's, which has created a Pavlovian connection in the human brain between tartar sauce and regret.

Or consider the monarch butterfly. The monarch's colouring is what's called aposematic, warning predators 'TOXIC! DO NOT EAT!' In a world of austere monochrome, the monarch has no warning signal with which to deter hungry frogs and birds. Instead its only hope of survival is to blend in with all the other graphite butterflies, slate-grey macaws, and pumice parakeets flitting through electrostatic skies that fizz like out-of-tune TV sets. Safely camouflaged in this way, the monochrome monarch may survive, but I fear for the scarlet kingsnake and all other practitioners of Batesian mimicry.

Batesian mimicry is a twist on aposematic colouring, by which edible snakes and frogs deter predators by mimicking toxic snakes and frogs. The edible non-toxic scarlet kingsnake, for example, copies the patterns and colours of the poisonous and venomous coral snake. The kingsnake's life depends on falcons, weasels and monitor lizards falling for the bluff.

'What these astounding phenomena teach,' the great art historian E. H. Gombrich wrote,

is precisely that there is a limit to perceptual relativism. What looks like a leaf to modern European must also have looked like a leaf to predators in fairly distant geological epochs. Likeness is not only in the beholder's eye.[*]

But colour, for Eagleman, is only in the eye of the beholder. 'Colour,' he says, 'is an interpretation of wavelengths, one that only exists internally.'

How could there be colours in our head if there were none in the world? From where would we get the concept? And why is colour the illusion rather than, say, colour-blindness? Why is the one supposed to happen only in the head but not the other?

In 1894 Arthur König demonstrated the fovea to be blue-

---

[*]   E. H. Gombrich, *Art and Illusion,* 2002.

blind. Does this mean the fovea is one step closer to seeing reality-as-it-really-is than the rest of the eye? Only the fovea has been able to rid itself of the blue delusion. It alone has escaped the shackles of blue to see the sky for the fizzing Alka-Seltzer electrostatic it really is. Now we just need to figure out how to communicate the fovea's disillusion to the rest of the eyeball.

The fact that the human brain perceives only one ten-trillionth of the spectrum of electromagnetic radiation is evidence enough for Eagleman to declare that 'in the outside world colour doesn't actually exist.' Shouldn't the fact that we can only see a narrow band of broader spectrum, suggest that the world is *more* colourful than we can possibly imagine, not less? That would be the logical conclusion, wouldn't it? And it would be consistent with zoology, too.

In 2016 it was discovered that reindeer not only see ultraviolet light but, in stark contrast to other mammals, their eyes have evolved to resist the damage caused by UV. This resistance allows them to spend longer staring at snow for any clues of food or foe, without becoming snow blind. It enables them to do better than humans or pine martens at discerning patches of hollow snow, which will give way if stepped on.

Reflectance spectrophotometry has revealed that blue tits ought really to be called ultraviolet tits, since females prefer males with the most dazzlingly UV crests. All this dazzle passes us by. Human eyes can't see it, but blue tits are not making it up. Their eyes just have different cones to ours, cones which are 'visually sensitive to wavelengths in the near-ultraviolet.*

---

* S. Hunt *et. al.*, 'Blue tits are ultraviolet tits', *Proc. Biol. Soc.*, 1998.

We cannot hear the high-end kilohertz laughter of tickled rats, nor the deep clicking of long-tusked narwhals echolocating their way through the black depths of the Arctic Ocean.

All of which goes to show only that there are biological constraints to what different animals can discover in their environments. Eagleman himself puts it very clearly when says that 'each creature picks up on its own slice of reality.' That is absolutely right. But it is not very melodramatic or spooky. It doesn't have the macho tone of 'can you handle reality as it really is?' And so Eagleman cannot stop there, but goes on to commit himself to the disastrous doctrine that animals have no access to reality at all:

> In the blind and deaf world of the tick, the signals it detects from its environment are temperature and body-odour … No one is having an experience of the objective reality that really exists.

If 'no-one is having an experience of the objective reality that really exists', then what is this temperature and body-odour that the tick thinks it detects? And didn't Eagleman just now say that the world was odourless? Or did I only imagine he said that. Did you imagine it too? If we both imagined it then maybe it is an objective reality that really exists. If so, then how can the deaf-blind tick *detect* body-odour? To detect means 'to discover or identify the presence or existence of.' But in an odourless world odour has no presence or existence. You can't detect odour in an odourless world.

Then there's the temperature the tick thinks he feels. If feeling cold is not 'an experience of the objective reality that really exists', the coldness the tick thinks it detects, therefore, is a crea-tion of itself and not out in the world. This begs the question: if icebergs, snow or permafrost are not in and of themselves cold, then how do they form?

I would like to propose a compromise. What about this? I believe the tick's perception of coldness might happily coincide with there being some actual coldness out there. I hope you agree with me. But, I'm afraid that if we want to stay true to Eagleman, this happy compromise is, alas, quite out of the ques-tion. His stern philosophy does not, we shall see, allow even this.

'The real world is not full of rich sensory events,' writes Eagleman. 'Instead our brains light up the world with their own sensuality.'

If the real world is not full of rich sensory events then why do animals suffer so badly from sensory deprivation?

In the 1960s at the University of California, Mark Rosen-zweig and Michael Renner showed that if you take two rat pups from the same litter, give them the exact same diet, same light, same warmth, but raise one in a bare cage and the other in a cage with running wheel, rope walk, mud, junk rubble and – best of all – other rats, then by simply comparing the two brains in autopsy, you can tell which rat grew up in a world full of rich sensory events and which did not. The brain of the rat raised in the impoverished conditions of a bare cage will have 25 per cent fewer synapses. Its cerebral cortex will measure up to 7 per cent thinner. There will be less capillary vasculation, and less dendritic arbourisation, unlike the rich bowers of dendrites all budding with fresh synapses observable in the rat raised in enriched conditions.

A lack of complexity in physical surroundings and social interactions leads to a lack of complexity in synaptic connections. Autopsies deduce the stunted conditions of a rat's life from the stunted brain. The proof that animals have access to the world outside their heads, therefore, is found inside their heads! The outside world, however imperfectly we perceive it, lights up our brain.

Environmental complexity has since been found critical for children between birth and six. Never again will your brain create so many new brain cells and new connections between them as it does in your first six years of life. (After that you've peaked). But for the brain to proliferate wildly, toddlers and young children need complex environments to play in. If not they will never fulfil the 'exuberant synaptogenesis' that is their birthright. Complex public spaces are especially critical if the child lives in a small, homogenous box surrounded by other small homogenous boxes. And so, in one of those weird and wonderful connections between totally different worlds, what was discovered in those Californian laboratories in the 1960s influences the design of inner city playgrounds to this very day. Thanks to those Californian experiments, the London Borough of Camden now makes sure that all its playgrounds include rope walks, rubble, mud, junk and rats.

Bishop Berkeley

Those discoveries about the effect of environmental enrichment on the brain were made at the University of California at Berkeley, a city named after the most famous proponent of the idea that the outside world cannot be known.

'Colours, sounds, taste,' wrote Bishop George Berkeley (1685–1753), 'have certainly no existence without the mind.'

Bishop Berkeley was very upset by Newton's Optics, and he responded with a series of anti-Newtonian tracts. Starting with *An Essay Towards a New Theory of Vision* (1709), he then broadened his attack in successive treatises and essays to include Enlightenment empiricists and materialists more generally. His philosophy, says Isaiah Berlin, is 'rooted in a pre-Renaissance medieval spiritualism,'* and yet it is this philosophy which makes a comeback in modern brain science books.

In *A Treatise Concerning The Principles of Human Knowledge,* Berkeley writes:

> It is indeed an opinion strangely prevailing amongst men that houses, mountains, rivers ... have an existence natural or real, distinct from their being perceived by an understanding.

To insist on the independent existence of houses, mountains, rivers and every last particle of matter 'must needs be a very precarious opinion; since it is to propose without any reason at all, that God has created innumerable beings that are entirely useless, and that serve no purpose.'

Though Berkeley wrote 'in Opposition to Sceptics and Atheists', he infuriated his fellow Christians just as much. Boswell tells us how he sent Samuel Johnson half-barmy:

> After we came out of the church, we stood talking for some time together of Bishop Berkeley's ingenious sophistry to

---

* Isaiah Berlin, *Age of Enlightenment*, 1984.

prove the nonexistence of matter, and that everything in the universe is merely ideal. I observed that though we are satisfied his doctrine is not true, it is impossible to refute it. I never shall forget the alacrity with which Johnson answered, striking his foot with mighty force against a large stone, till he rebounded from it – 'I refute it thus."

I love how personally Johnson takes this. When the most eloquent Englishman who ever lived kicks a rock in fury he reminds us that some propositions are best not answered in cold blood. To deny someone any claim to any kind of contact with reality, as Berkeley does, is an act of psychological violence. It is the weapon of bullies, the tactic of hostile interrogators who try to browbeat and bamboozle a private soldier out of making a complaint against senior officer. You didn't see what you think you saw. They didn't say what you think you heard them say. You weren't even where you think you were when you saw what you thought you saw.

Why is there nothing not something?

For thousands of years one of the fundamental philosophical questions has been why is there something not nothing? With Eagleman we find ourselves in the strange position of asking why is there nothing not something? Why would the objects of the world have no texture, no taste, no sound, no smell – rather than something, anything, even if different from what we think?

---

* James Boswell, *The Life of Johnson*, 1791.

Why none at all? I could understand if he was saying everything was mauve, had the texture of tulip petals, and the taste of ash, but why sans taste, sans everything? This goes far beyond Berkeley, for whom the things of the earth, even though they depend on being perceived for their existence, are eternally real because forever under God's good gaze.

To answer to the question 'why is there nothing not something in Eagleman's philosophy?' we need to look at what is real for him. What is still standing once he has razed the outside world? And the answer, it turns out, is: *wavelength frequencies*.

The blueness of a Japanese lizard's tail is an illusion entertained by the weasel, the snake and me. What is not an illusion, however, what is in fact irrefutable is the electronvolt energy value of the light bouncing off its tail. Why are wavelengths true but not a lizard's bright blue tail? It is, I suggest, because we have left the real word of science for the virtual world of Neuropolis, where, inscribed above the city gates, is that great motto of scientism:

All science is either physics or stamp-collecting.

But Ernest Rutherford's boorish remark is false. All science is not physics. If you want to find out how lizards are tricking weasels into attacking tails not heads, an isotopic triaxial probe is simply the wrong tool for the job, because the job isn't about measuring electromagnetic frequency. The job is ecological, and 'ecological events must be distinguished from microphysical and astronomical events.'[*]

---

[*] James Gibson, *The Ecological Theory of Visual Perception*, 1986.

It's a question of scale, as much as anything else. Microphysics might accurately describe a stream as atoms colliding, or wavelengths oscillating, but when we wade barefoot across the stream our experience isn't an experience of atoms and electromagnetic wavelengths. We experience wetness and cold. The stream's pebbles are treacherously slippy, with a sort of slime on them, and wedge the feet bones apart in a surprisingly painful way.

Are these merely subjective impressions when what science demands are objective measurements? Not if science demands an accurate description of animal interacting with environment. If that's what we want then an accurate description must be at the ecological level. That is the appropriate level for the job in hand, since our lives are lived at the ecological scale – not among the celestial objects of astronomy or the neutrons of the microphysical realm, 'but in the very world', as Wordsworth wrote,

> which is the world
> Of all of us, – the place where in the end
> We find our happiness, or not at all!*

Austerity on the brain

For David Eagleman, austerity is deeply woven into the fabric of nature. It is not an invention of humans, as he believes colour to be, but intrinsic to matter, to reality. Not only is nature austere as in grey and dour (a claim we examined earlier) nature is also

---

* William Wordsworth, 'The French Revolution as It Appeared to Enthusiasts at Its Commencement', *The Prelude*, 1805.

austere as in pinched, frugal, economizing. Eagleman doesn't apply austerity measures to the living world, he just discovers that the living world proceeds according to austerity principles. It turns out that organs such as the brain, for example, conduct a thorough review of all non-essential services:

> So why doesn't the brain give us the full picture? Because brains are expensive energy-wise… brains try to operate in the most energy-efficient way possible.

I know energy-efficiency would seem to be something you might expect from a clever organ like the brain, but, for better or worse, that appears not to be the case. Whereas a smart electrical appliance, for example, powers down when not being used, our brains are more active when we sleep.

One of the most wonderful features about how the brain works, in fact, is the sheer extravagance of neural activity, its superabundance. The technical term used to describe the synaptic proliferation that characterises early brain development is 'exuberant synaptogenesis'. In the landmark paper 'The Physiology of Perception', Walter J. Freeman and his colleagues found that 'perception depends on the simultaneous, cooperative activity of millions of neurons spread throughout expanses of the cortex.'[*]

Not exactly a slimmed-down organization. A rationaliser seeking ambitious saving targets would ruthlessly downsize such a sprawling operation, and would also take the axe to this sort of spare capacity:

---

[*] Walter J. Freeman, 'The Physiology of Perception', *Scientific American*, 1991.

vast collections of neurons ... shift abruptly from one complex activity pattern to another in response to the smallest of inputs.[*]

We are told every day that public sector social services should be streamlined. This is dunned into us with such monotony that it begins to look like a Law o' Nature, rather than one political choice among many other possible ones. Defunct economic dogma does not apply to how the brain works. Whatever the political and economic weather the brain continues its extraordinarily successful policy of being extremely unstreamlined. Just take a look at the Spanish practices going on in entorhinal cortex.

The entorhinal is famous for spatial navigation and memory. Two paths – lateral and medial – lead from the entorhinal to Memory Central in the hippocampus. The lateral path is for spatial navigation 'Where am I?' and the medial for memory 'What happened?' Management consultants, who make it their business to rationalise a firm to its knees, have a horror of what they call 'duplication of function', but I'm afraid that's what we have here. In a regrettable recidivism, wholly ignorant of best practice guidelines (helpfully supplied by Goldman Sachs) and the harsh new economic realities (also helpfully supplied by Goldman Sachs) the brain simply refuses to 'operate in the most energy-efficient way possible'. I blame the unions.

The only time the entorhinal cortex is not guilty of 'duplication of function' is when it is busily triplicating. It's not enough for the entorhinal cortex just to check sense data from the hands

---

[*] 'Walter J. Freeman, 'The Physiology of Perception', *Scientific American*, 1991.

against data from the eyes, it also insists on cross-checking with the middle ear, in a process called reentrant mapping. Here we have unforgivable 'triplication of function'. Yet it all works very well, and has done since before records began.

Now it is true that bodies need to conserve energy. But the reason the brain doesn't give us the full picture has little to do with the brain being anxious about squandering the energy budget all in one go. (After all, the brain never seems bothered by wasting its 20 per cent share of our energy budget watching three-minute clips of *The Sopranos* on YouTube for five hours straight.) The reason the brain doesn't give us the full picture is not to do with making energy savings, but because it has evolved to privilege motor activity above all else.

We need to act in real time. We need to do things now. We are surrounded by predators and prey, many of whom come equipped with vastly quicker reflexes than our own.

When *Homo ergaster* is sprinting to grab her toddler, she doesn't need to know whether the puma is male or female, or in fact a jaguar. For now, Big Cat Prowling will do. Once you have your toddler in your arms, once your shouts and screams have brought stone-throwing elders to your aid, once the big cat is at bay, then and only then is it useful to notice second-order facts: that the puma is arthritic or old, or that is only a much less scary lynx or linsang. But in those split seconds of your initial reaction all you need to know is where and what in the roughest possible sense. Your picture does not at first need to be more detailed than that – in fact, more detail would not help but hinder.

A good illustration of this is the story of why, during the World War II, my Auntie Ada was discharged from the

Women's Auxiliary Air Force (WAAF). From 1940–42, Ada Newman, aged 22, worked in the map room of the RAF Group Operations HQ located in a secret bunker beneath the Strand in central London. Her job was to push model tanks and planes across a giant horizontal map with a croupier stick in response to grid references being called out by WAAFs on headphones. The map was enormous, the size of four table-tennis tables stuck together. It was a relief map with models of forests, mountain ranges, and painted streams and roads. When pushing a model tank through the model forest, or landing a model plane on the shore, Aunt Ada used to do engine noises and gunfire sounds.

Superiors gave her verbal warnings but she couldn't help herself. She said she didn't know she was doing it. The bending end came when Air Chief Marshal Hugh Dowding came to the map room with senior figures from the Admiralty on the eve of a joint sea-air attack. While Aunt Ada was moving a line of German infantry, the top brass overheard her saying:

Gott in Himmel! Once again Tommy's air cover has proved superior to our anti-aircraft fire. But I die … for … ze … fatherla—aannnggh!

Minutes later, while moving an aircraft carrier from the open sea of grid reference A1 to the harbour of E3, it seems Ada Newman found herself doing a ship horn, followed by the sort of nautical rhubarb the Beatles get up to on *Yellow Submarine*. She was relieved of map room duties and then discharged from service. In her dismissal hearings, she claimed that she'd only been trying to concentrate the minds of top brass by giving them a more vivid picture of reality down on the ground. In response to this, her superior officer read out the transcript of what Auntie Ada had said:

'Aye, Aye Cap'n. Full Speed Ahead. Steady As She Goes. Hard To Port.' He then looked up from the transcript, and in a voice that Auntie Ada said was unnecessarily harsh, asked, 'In what possible way is that concentrating minds?'

Family feeling aside, I suppose the superior officer had a point. The tanks in the wartime map room are not supposed ot be detailed or individuated. Their role is to give us a big picture at a glance so as to enable a rapid response. It's the same with the brain in an emergency. The brain doesn't give us the full picture straight away is because it has evolved to serve action in real time, like the fox in Ted Hughes' poem *The Thought Fox*:

Two eyes serve a movement, that now
And again now, and now, and now.

Sets neat prints into the snow.

It's bad enough that we have to endure a fake and made-up economic austerity, without having to accept an equally fabricated natural austerity. What makes this austere explanation of vision so very galling is that it comes in the middle of a hyper-inflationary bonanza of unregulated speculation that the world is in actual fact, silent and monochrome, and reality takes 'place in the sealed auditorium of the cranium' and all the rest of it.

'Practical men', wrote John Maynard Keynes, the most influential economist of the twentieth century, 'who believe themselves to be quite exempt from any intellectual influence, are usually the slaves of some defunct economist.'

I think this great insight is true of those practical men who write neuroscience books full of handy tips, such as 'the brain is a tool-kit.'

The most influential philosophers today are people who wouldn't dream of calling themselves philosophers. They write books and make TV series about who we are and where we came from. They claim merely to extrapolate from what the science says. But, to paraphrase Keynes, those who believe themselves to be dispassionately reporting what the science says are usually the slaves of some defunct philosophy.

When we think we are most free from philosophy, we are most under its spell. If we are not aware of where ideas come from, then it's harder to resist their influence. But by tracing the sources back to the philosophical stowaways, we may better glimpse ways to escape a deadening neuro-mythology.

The 'Book of Joshua' (in which God commands the earth to stand still) once stood in the way of understanding planetary motion. Today, the austerity model stops us understanding the first thing about how the brain works. Eagleman is standing in front of the stage filming the gig on his phone, and blocking our view of the brain's funky moves.

4. WHEN YOU'RE SNARLING

In *Phantoms in the Brain*, neuroscientist V. S. Ramachandran, who is listed by *Time* magazine as one of the 100 Most Influential Thinkers in the World, speculates on the evolutionary origins of smiling. Smiling, he says, evolved from an aborted snarl. He bases this theory on no evidence. Instead he advances the following fantasy scenario:

> When one of your ancestral primates encountered an individual coming towards him—

And right there, by the way, isn't that a curious and rather telling choice of phrase? Not 'one of our ancestral primates', but one of *yours*. Clearly Ramachandran is cut from superior cloth. We may have come down from the trees, but he descended from the mezzanine on a spiral staircase to proclaim:

> When one of your ancestral primates encountered an individual coming towards him, he would have bared his canines in a threatening gesture on the fair assumption that most strangers

are potential enemies. Upon recognizing the individual as friend or kin, however, he might abort the threatening grimace halfway, thereby producing a smile, which evolved into a ritualized human greeting.

It is desperately sad that someone could look around the world in which we live, and in every expression of joy, gladness, fellow-feeling and goodwill see only a snarl, like some kind of upside-down, inside-out, back-to-front Louis Armstrong.

I see friends shaking hands,
Saying 'how do you do?'
They're really saying,
You stupid bastard look at you in your fucking tassel loafers, you
ghgnnrghfcckgg!'

Well, you may not like Ramachandran's conclusion, Newman. It may upset your happy-clappy, eco-hippy worldview, but it just so happens to be what the science says …

Well, it's certainly not what comparative anatomy tells us. Because if you want to talk about ancestral primates, it so happens that when chimpanzees laugh top lip covers upper teeth. Baring the canines doesn't come into it. When chimps laugh they expose only their lower teeth while swinging their jaw form side to side. (How do I know this? Let's just say I don't always play my first choice of venue.)

But don't take my word for it. Charles Darwin noticed the same thing in the course of his own investigations into the evolutionary origins of smiling. In *The Expression of the Emotions*

In Man and Animals, Darwin comes to a very different conclusion about the evolutionary origins of the human smile:

> Our long habit of uttering loud reiterated sounds from a sense of pleasure [has evolved] into the tendency to contract the orbicular and zygomatic muscles whenever any cause excites in us a feeling which, if stronger, would have led to laughter.

So where Ramachandran says that a smile is halfway from a snarl, Darwin says that smiling is – to quote a recent review of my standup – halfway to a laugh.

If Ramachandran's fantasy about the evolutionary origins of smiling doesn't come from comparative anatomy, ethology, zoology or evolutionary biology, then where does it come from?

It has its roots, I believe, in that most tenacious of philosophical stowaways, Romanticism.

'Laughter', wrote mid-nineteenth-century French Romantic poet Charles Baudelaire, 'is a man's way of baring his fangs.' Baudelaire's idea of comedy, commented Jean-Paul Sartre, 'is entirely of a piece with his frigidity, sterility, and complete lack of empathy.'

Sounds like a description of each and every privately-educated stand up comedian now dominating our cultural landscape, all doing variations on the same theme: 'Is it me, or is everything shit? Have you noticed? I mean, is it me? Or is everything shit? Is it me? Or is everything shit?'

It's both, and there's a connection.

Now, I don't want the fact that I have just quoted Jean-Paul Sartre to be taken as any kind of endorsement of either him or the Existentialists. Not least because Existentialism is partly to blame for such bizarre notions as We Are Our Brains – which entails that our bodies are somehow not really part of us, as if we are not part of nature at all, just isolated entities floating around, too good for this earth, not really belonging here, yet not really belonging anywhere else either. It's this cluster of fallacies that is going through Jean-Paul Sartre's head when he talks about – and I love this quote – 'the nauseating sloppiness of the natural world'. Here is a man, you feel, who when he listens to the dawn chorus says, 'One at a time!'

Sartre displays here only his own nauseatingly sloppy thinking about the natural world, which is, of course, full of animals doing finely calibrated, precision engineering and detailed painstaking work.

Consider *Cyclosa tremula*, a black and white striped Guyanan spider. *Cyclosa* builds replica spiders with which she populates her web. Because she makes these replicas out of prey debris, the husks of insects she has devoured, they do not have her vivid black and white stripes but are a dull grey colour, which the local birds soon learn not to bother eating. And so, when *Cyclosa* sees an orange-bellied sparrow swooping overhead she bounces up and down on her web, blurring her lines, blending black and white to make grey and the sparrow flies away. It seems, then, that *Cyclosa* is building her replica spiders as a cunning decoy, in much the same way as during the World War II the British Army built thousands of dummy cardboard tanks. Their turret guns were made from the long cardboard tubes inside wallpaper rolls. They were painted in green and brown camouflage colours and dotted around the fields of Kent alongside hundreds of full-size *papier-mâche* Spitfires. There were even two entire dummy barracks made from crepe paper stretched between willow hurdles with a felt roof. The reason for all this activity was because we knew that German spy planes were flying overhead taking photographs. The hope was that when these photographs were developed in Berlin, the German High Command would take one look at them and say: 'There's no point trying to invade Britain, the whole country's made out of craft materials.'

Smiling was big in the Enlightenment. Humphry Davy and Thomas Paine, Diderot and d'Alembert can be seen smiling in their portraits. Julien Offray de la Mettrie is not only grinning but also wearing what appears to be a shower cap. To the sculptor Houdon, the twinkle in Voltaire's eye was so intrinsic that it inspired a piece of witty improvisation when he came to do a bust of him. Houdon cemented a tiny stone onto each eye of his Voltaire bust. You'd think nothing could be further from a glint in the eye than a bit of gravel, but amazingly it works. Most people see the twinkle and not the grit of which it is made. And then of course there's the famous oil painting of Volatire used on the cover of the Penguin Classics edition of *Candide*, bare-legged under his white cotton nightdress, grinning from ear to ear.

Come the Romantics, however, and smiling was out. In Paris's new photographic studios, Baudelaire was determined to make a decisive break with those grinning Enlightenment loons. No-one was going to catch him smiling at the birdie as the flash-gun whoompffed. No way. Not wanting to smile for the photographer or the oil painter doesn't automatically make someone a poseur of course. Perhaps Baudelaire was more aware of life's ugly reality than the Enlightenment optimists. Perhaps he had seen the world for what it really was, and not spent his life hopping excitedly around an air pump. Perhaps that was why he saw the fang beneath the smile. That's certainly the pose Baudelaire strikes in his most famous poem *Flowers of Evil* (a title, you feel, that even Iron Maiden would reject), but there is a problem with the argument that Baudelaire is pessimistic because he has been around the block a few more times than those shallow Enlightenment optimists. The problem is that cock-a-hoop Julien *'Is it all right if I keep my shower cap on for the picture?'* de la Mettrie was a battlefield surgeon in the War of Austrian Succession, whereas Baudelaire lived with his mum.

As did Jean-Paul Sartre – for forty years! – which I think goes a long way in explaining the macho tone of his philosophising. If you are afraid of the dark, he says, it is because you *choose* to be afraid of the dark. Not Sartre. His last words every night, after being tucked up in bed, were: *'Eteins la lumiere, Maman. J'ai choisi une vie sans peur!'*

'Are you sure, *mon petit*? Shall I leave the landing light on in case you don't make it to the toilet in time and have a little 'accident' again like last night?'

'I chose to wet myself. I enjoy the sensation of wet pyjama.'

'Not in my house you don't. When you move out and get a place of your own, you can piss all over it to your heart's content, Jean-Paul, but as long as you live here the landing light stays on. And that's final!'

The Romantics and Baudelaire were by no means the first people ever to entertain strange ideas about smiling and laughter. For seventeenth-century philosopher Thomas Hobbes, for instance, laughter came from a sense of 'sudden glory arising from a feeling of superiority.' Well, that is one sort of laugh. But do we crow like that when our child takes her first steps? Surely the laugh we laugh then comes from delight, not from a sudden sense of how immeasurably better at walking we are than her? We tend not to snap our fingers in her face, and say, 'Call that walking? Hah! Oh dearey me! Woeful! Oh my sides!'

Thomas Hobbes enjoyed the tremendous luck of getting to hang out with both Galileo and Ben Jonson. You'd think somewhere between Padua and Eastcheap, Hobbes might have noticed different kinds of laugh than the triumphant snort.

So, there is a long history of curious ideas about smiling and laughing, yet none so strange as our new notion that smiling is an aborted snarl. Since Ramachandran is one of the 100 Most Influential Thinkers in the World Today, then it is worth paying close attention to the specifically neuroscientific roots of this extraordinary notion.

To this end we must unpick a tangled web that includes a famous industrial accident, a Gothic melodrama about a crazy scientist, and Bumpology, 'The One True Science of The Mind.'

5. PHINEAS GAGE AND THE MYTH OF THE SUPERMAX BRAIN

In 1848, Vermont, USA, railroad worker Phineas Gage was tamping an explosive charge into a pre-drilled hole in a granite rock face in order to blast a cutting for the Rutland & Burlington Railroad. Sparks from his tamping iron set off the explosive charge, firing the 13lb iron bar clean through his skull, taking out what we would now call his pre-frontal cortex.

Phineas Gage survived the accident to become an instant medical celebrity. When distinguished Harvard physicians came to study him, they were amazed to discover that he appeared to have suffered no mental impairment whatsoever. Except one. He was no longer able to behave in a socially appropriate way. He'd have sudden fits of rage characterised by the use of what the visiting Harvard physicians called 'grossest profanity'. And that, say the textbooks brightly, is how we know that the pre-frontal cortex is the bit of the brain responsible for self-control, and for mediating socially appropriate behaviour.

But nobody's looking at this from Phineas Gage's point of view! If I was him, I'd be saying:

I've been listening to all you eminent physicians puzzling over what could possibly be causing my wild mood swings, and my regrettable slide into the use of gross profanity, and you know what's just crossed my mind? An iron fucking bar. Now if a man cannot cuss when four feet of metal rod shish-kebabs his brain, when can he cuss? Trust me, when this happens to you, tarnation is not the word you are looking for. A darn won't do you now. This is no Jumping Jehosaphat type of situ-fucking-ation. I nearly DIED!!!

His life was saved by first responder Dr Edward Williams, who found Phineas sitting on his porch fully conscious despite the hole in his head, from which he removed, as he later wrote, coagulated blood, shards of splintered skull and 'approximately three ounces of brain material'.

Now what I want to know is how do you know when to stop taking the brain material out? I guess Dr Williams scooped out an ounce at a time. There's Phineas sitting on the porch, and Dr Williams is standing over him with a tablespoon.

Dr Williams: I can see some loose and flappy bits of brain in there, Phineas, that are gonna have to come out. Now I'm gonna scoop out an ounce at a time. If at any point it feels sketchy, you just holler and I will immediately desist. Okay. First ounce coming out now... Hup! How was that?
Phineas: Didn't feel a thing, Doc.
Dr Williams: Right, Phineas. Here we go. Second ounce – hup! – out it comes! Okay?
Phineas: Can't say I feel any different at all, Doc. You go right ahead.

Dr Williams: Okay third ounce. Third ounce coming out now, hup! How's that?

Phineas: I think the British people will welcome a state visit from President Trump.

Dr Williams: Gotta put that third ounce back. That's the soul right there! That's what separates us from the baboon. You need that third ounce!

Almost all brain science books tell the Phineas Gage story. But it is strange that those who claim to be experts on how the mind works should be unable to grasp that this young man's state of mind might be down not just to his tattered brain but to what he thinks and feels about his tattered brain. Neuroscientific accounts never entertain the possibility that Phineas's rage might be due to grief or shock or even simple pain from his shattered jaw and eye-socket. Instead, they tell the story of Phineas Gage as illustrating a sort of Dr Jekyll and Mr Hyde model of the human brain. There's the snarling Mr Hyde, our animal self, the killer ape inside, the product of millions of years of evolution, the real us, barely restrained by Dr Jekyll, the late cortical add-on, product of a few thousand years of flimsy social contract. The iron bar that shoots through Phineas Gage's skull rips a hole in this cortical crust allowing the sociopathic Mr Hyde to escape Dr Jekyll. Strange to say this has become the standard scientific model in all neuroscience textbooks, the curtain-raiser on the study of cortical localisation, the science of which bits of the brain do what. What is especially strange about the acceptance of this melodramatic version of events is that it is totally un-Darwinian.

For Darwin, brain trauma doesn't reveal our true animal nature, *it separates us from our true animal nature.* For Darwin, as we have seen, snarling is no more atavistic than smiling, aggression no more human than sociability. 'We have every reason to believe,' argues Karl Popper in the same vein, that our ancestors 'were social prior to becoming human'.*

The Myth of the Supermax Brain

In the struggle for the survival of ideas, Robert Louis Stephenson's fiction is selected over Darwinian fact. Jekyll and Hyde better fits the modern Myth of the Supermax Brain.

According to this myth, the prefrontal cortex operates like a supermax prison locking down the seething violent criminality of our true selves.

The Supermax Myth is popular because it ticks so very many boxes about how the mind ought to work. Here it seems is the bridge between Freudian psychology and modern neurobiology, between the psychoanalyst's couch and functional magnetic resonance imaging. Such a strategically important bridge is always going to be defended with ferocity. Only fanatical loyalty to the Supermax Myth can, I think, explain V. S. Ramachandran's curious hostility towards Phineas Gage. In *Phantoms in the Brain*, he tells us that after the accident Gage became 'a worthless vagabond with absolutely no moral sense'.

The asperity is startling, not least because it flies in the face of the historical record. We know that the worthless vagabond

---

* Karl Popper, *Conjectures and Refutations: The Growth of Scientific Knowledge*, 1962.

continued to support his family, working on their smallhold-ing in Enfield, New Hampshire. In the remaining thirteen years of his life, he took on a series of increasingly demeaning jobs, despite suffering seizures, blackouts and terrible headaches.

In what follows I am indebted to Malcolm Macmillan's painstaking research into first-hand sources, archive material and contemporary witness statements as he single-handedly disinterred man from myth in his book *An Odd Kind Of Fame: Stories of Phineas Gage.*

One hot and dusty day in August, 1849, Dr John Jackson trav-elled from Boston to Enfield, New Hampshire to interview the Gage family. He'd been hoping to examine Phineas himself, but met only his widowed mother and brother-in-law, who told him Phineas was in Montpelier trying to get work with another rail-road company 'doing what he did before'.

I confess that I had to re-read that last phrase three times over, when I first came across it in *An Odd Kind Of Fame.* Doing what he did before …? Astonishingly, Phineas Gage was trying to find work as a blasting foreman! I guess he was hoping to impress the Montpelier railway company with his experience more than his skill. Then again, who better than he to instruct railway navvies on how really, really careful you should be when priming an explosive with your tamping iron? It is after all a moot point whether we listen more attentively to the one-armed or two-armed bomb disposal expert.

Dr Jackson stayed to interview Phineas's mother Phebe Gage, still in black crepe since the death of her husband a few months earlier, and made notes of their conversation. Jackson began by asking Phebe Gage about her son's recuperation, and jotted down her reply:

> abt. February he was able to do a little work abt. ye horses & barn, feedg. ye cattle &c.; that as ye time for ploughing came he was able to do half a days work after that and bore it well.[*]

Dr Jackson then asked after her son's mental state. She replied that for the first few days after the accident her son was 'childish', but now he was back to his old self – except his memory was impaired. 'A stranger would notice nothing peculiar', she told Dr Jackson, but she did and so did the rest of the family.

Clearly, Phineas Gage after the accident was not the same man as before. Brain damage changed who he was, but did it extinguish who he was? Did it reveal for our edification some ancestral primate? Not for Phineas's family at least. They saw in him the same industrious young man he had always been, eager to get on, and so impatient to be well again that he even ploughed a field before he was fully recovered. He was also, it seems, anxious to retain the hard-won status of blasting foreman, even if it meant he had to travel the sixty-five miles from Enfield to Montpelier in hopes of finding a firm who would hire him despite his disfigurement and, uh, track record.

This, then, is the raving wild man of neuroscientific myth, the worthless vagabond with absolutely no moral sense. Ramachandran's belief in a mythology unsupported by evolutionary biology commits him to a version of events unsupported by the historical

---

[*] Malcolm Macmillan, *An Odd Kind Of Fame: Stories of Phineas Gage*, 2000.

record. If Phineas Gage *isn't* a worthless vagabond, then we have to abandon the Supermax Myth, the bridge between psychology and neurobiology, and completely rethink our conception of the brain. At this point Ramachandran's acolytes helpfully suggest: 'An Open Prison, perhaps?' At which point one can only smile politely, tip one's hat and bid them each good day.

In her brilliant essay *Absence of Mind,* Marilynne Robinson singles out for attention the 'oddly stereotyped way' in which brain books handle the issue of Phineas Gage's swearing as if this somehow showed that the beast within had escaped. This is especially odd, she says, as what could be more human than swearing? So far as we know we are the only animals that do it.

That said, it will be a sad day if we finally decipher low frequency whale music only to discover that humpback whales are hurling long drawn-out expletives across the ocean at each other:

'Yyeeewwww waaaaaannnkkkaaaaahhhhh!'
'Yeeeeewww ffffuuuucckkkkkiiiinnggg bbaaaaasssstttttaaaaaaard!'

To complement her argument, I'd just like to add another reason why I think the neuroscientific literature's fixation on Gage's swearing is odd. Railway navvies were as famously foul-mouthed as mule skinners. If a man was swearing among a gang of navvies, who would notice?

In 1838, an engineer working on the London-Birmingham Railway, said that English navvies were:

> Possessed of all the daring recklessness of the Smuggler, without any of his redeeming qualities, their ferocious behaviour can only be equaled by the brutality of their language.

In *The Railway Navvies*, Terry Coleman puzzles over why there are not more reprints of Amercian navvy worksongs, and concludes that the songs were so sweary and blasphemous that they were 'considered unprintable and so were lost'.

And yet every telling of the Phineas Gage story says that his co-workers were shocked by his swearing, and always includes the following po-faced quote from a navvy: 'Gage isn't the same Gage anymore'. These Blushing Railroad Workers of Vermont come across like Monty Python's lumberjacks, who skip and jump and like to press wild flowers.

There is in fact a very good reason why contemporary observers attached great significance to Gage's swearing. But this dramatic significance is lost to us so long as we use anachronisms like orbitofrontal cortex, ventromedial frontal lobe, or pre-frontal cortex. What the iron bar destroyed was not the ventromedial frontal lobe, but the Organ of Veneration, for the 1850s were the heyday of phrenology.

The Organ of Veneration

If we keep in mind that Phineas Gage's Organ of Veneration has been destroyed, then the focus on his gross profanity begins to make sense. It wasn't the swearing that got everyone's attention, it was the swearing *in front of his betters*. No-one cares what oaths low people hurl at each other, but when ushered into the presence of someone venerable like Henry Bigelow, Professor of Surgery at Harvard Medical School, you keep a civil tongue in your head. To swear where you should venerate is a shocking abrogation of fundamental social norms, like a soldier patting his commanding officer on the bottom and saying, 'What's with all this ordering about, love? If you want to get on in this world, ask nicely.'

The Organ of Veneration is located front and centre on the phrenology chart, one of the largest single areas of the brain, a Spain to the Portugal of the Organ of Human Nature, for example. The Organ of Veneration's pride of place reflects a nineteenth century concern with hierarchy and rank. No veneration, no order.

In the nineteenth century, phrenology was not the quackery it later became, but the cutting edge of neuroscience. Phineas Gage's disaster allowed nineteenth-century medical science to refine the phrenological map. To this end, Dr John Harlow conducted an experiment to test the damage to his Organ of Comparison, which was close to the Organ of Veneration and as such lay in the tamping iron's flight path.

Finding Phineas playing catch with a handful of pebbles, Dr Harlow offered to buy four of the pebbles from him for one thousand dollars. Gage politely declined the offer, possibly fearing that the doctor had suffered irreversible damage to his Organ of Comparison.

For great apes dollar bills and pebbles are equally meaningless. The Organ of Comparison lay in the part of the forehead that apes do not have. Dr Harlow is literally seeing how far from the human state into apehood Phineas has fallen. He is playing a kind of Ker-plunk of the brain: how many phrenological sticks can be removed before humanity falls away completely and we have a bipedal ape? Phineas Gage offered a particularly good starting place for such an inquiry by dint of his profession, because even in full health, the navvy was popularly, if half-seriously, describe as the missing link between apes and humans:

> With fury and frenzy and fear,
> That his strength might endure for a span,
> From birth, through beer to bier,
> The link 'twixt the ape and the man.

'The Navvy Chorus', *Songs of a Navvy* (1912).

What did Dr Harlow make of Gage's waving away a grand? Did he take it to prove that, whatever else the tamping iron wrecked, at least Phineas's fundamental decency was still intact? Was this evidence that the accident hadn't damaged his patient's Organ of Conscientiousness (which was located halfway between the Organs of Sublimity and Firmness)? Alas, no. What Dr Harlow concluded was that Phineas Gage's refusal to trade gravel for dollars demonstrated an inability to compare worth and worthlessness, which therefore proved that the Organ of Comparison was destroyed by the tamping iron, which means it must be just where the phrenological map said it was. One of the crucial barriers between ape and man was down.

Even allowing for the fact that phrenology – or Bumpology as its detractors called it – was considered by advocates such as Dr Harlow to be The One True Science of the Mind, I find this a puzzling conclusion. If Phineas Gage doesn't know the meaning of money, if he thinks gravel is as good as gold, then why travel sixty-five miles to Montpelier for that job interview? Why get any kind of job at all, for that matter, when there's so much valuable gravel lying about all over the place, just there for the taking?

How I wish Phineas Gage had pocketed those ten green hundred-dollar bills, tapped the side of his nose, and said: 'And there's plenty more where these pebbles come from, Doc!' At that point Dr Harlow, blinking down at his palm and the four $250 dollar pebbles he now owned would say:

'Actually it was just a test, Phineas.'
'Say what?'
'Please can I have my money back?'
'You know, Doc, ever since the accident, if someone vexes me by like, fucking with my mind, I just go apeshit. Just lose it. Go fucking mental. I can't keep a lid on my temper any more cos now I ain't go no lid. Accident blew it off, know what I mean? I'm a fucking apeman, a wild man. So if someone like you was to, you know, say one thing and then the complete opposite? Well, let's just say I wouldn't be the only man in town with an iron bar in his head, you know what I mean? Now get off my property and take your mini fucking rockery with you!'

Nothing in the Phineas Gage story makes sense except in the light of phrenology, but phrenology is played down in popular retellings because Gage's accident is supposed to represent a decisive break with the past. With a big bang and a cloud of smoke the new science of cortical localisation is born. A couple of years later in 1861 Paul Broca publishes 'Sur le principe des localisations cérébrales', in the *Bulletin de la Société dAnthropologie*, in which he announces to the world how reason and emotion are divvied up in the brain:

The most noble cerebral faculties have their seat in the frontal convolutions, whereas the temporal, parietal and occipital lobe convolutions are appropriate for the feelings, penchants and passions.

Broca's schema betrays how both the new science of cortical localisation and the old science of bumpology share a common ancestor in the ancient Greek idea that Reason is a chariot-eer controlling the wild beasts of Passion. A line straight as a tamping iron runs from this Greek idea, through Broca rele-gating emotion to a penchant, and all the way to the Myth of The Supermax Brain. This tradition, I think, helps explain why Ramachandran bares his canines in such a ferocious snarl at the 'worthless vagabond with absolutely no moral sense.'

Incidentally, no-one ever accuses the Rutland and Burling-ton Railroad Company bosses of having absolutely no moral sense, even though they never paid Phineas Gage one red cent in compensation. But that's probably because rail bosses destitute of human decency were seen as just one more occupational hazard in the working life of a railway navvy, as this nineteenth-century American railroad song makes clear:

> *Last week a premature blast went off,*
> *A mile in the air went Big Jim Goff.*
> *When the next pay day came round*
> *Jim Goff a dollar short was found.*
> *When he asked what for, came this reply:*
> *'You're docked for the time you was up in the sky!'*

6. ROBOT NEW MAN

The benchmark for Artificial Intelligence (AI) is the famous Turing Test. Alan Turing's 1950's thought-experiment states that if a robot can convince you that you're talking to another human being, then that robot can be said to have passed the Turing Test, thereby proving that there is nothing special about the human brain that a sufficiently powerful computer couldn't do just as well.

Except the Turing Test proves no such thing. All it proves is that humans can be tricked, but everyone knew that already ... except Alan Turing, alas, who in the last week of his life – and this is a true story – went to a funfair fortune-teller on Blackpool promenade. Nobody knows what the Gypsy Queen told him, but he emerged from her tent white as a sheet and killed himself two days later. But funfairs have had centuries of practice in the art of tricking punters.

Weirdly, a funfair nearly did for Isaac Newton. In a posthumous biographical sketch, his friend John Wickens says that when they went to Sturbridge County Fair, Newton had a complete meltdown, and was close to jettisoning his whole theory of how gravity acts on every object in the universe, after what Wickens describes as: 'a frustrating hour at the coconut shy'.

In an interview with *The Times* about Artificial Intelligence, Brian Cox said:

> There is nothing special about human brains. They operate according to the laws of physics. With a sufficiently complex computer, I don't see any reason why you couldn't build AI. We'll soon have robot co-workers, the difference is we'll even be taking them to the office party.

I wrote a letter to *The Times*. They didn't print it. I don't why. It was quite short. It just said: 'No we fucking won't'.

Emotional robots are a vision of the future to be found in the Gypsy Queen's crystal ball but not in science. Not least because of these two uncontroversial scientific facts:

1. We are not machines, we are animals.
2. No experiment performed by anyone anywhere in the whole world at any time has found a shred of evidence to suggest the remotest possibility that a 'sufficiently complex computer' will ever be able to do literally the first thing that a mammalian brain does, and experience emotion.

We came crying hither.
Thou know'st the first time that we smell the air
We wawl and cry ...

But to listen to AI cultists you'd think we were knee-deep in this sort of evidence. According to Radio 4's *Inside Science* program, for example, we'll soon have robot lawyers.

A senior IBM executive explained to *Inside Science* listeners that while robots can't do the fiddly manual jobs of gardeners or janitors, they can easily do all that lawyers do, and will soon make human lawyers redundant.

Interestingly, however, when IBM Vice President Bob Moffat was himself on trial in the Manhattan Federal Court, accused in 2010 of the largest hedge-fund insider trading in history, he hired one of those old-time humanoid defence attorneys. A robot lawyer may have saved him from being found guilty of two counts of conspiracy and fraud, but when push came to shove, the IBM VP knew there's no justice in automated law.

Not all the gigabytes in the world will ever make a set of algorithms a fair trial. There can be no justice in the broad sense without procedural justice in the narrow sense. Even if the outcome of a jury trial is identical to the outcome of an auto-mated trial, due process leaves one verdict just and the other unjust. Justice entails being judged by flesh and blood citizens in a fair process. Not least because victims increasingly demand that the court consider their psychological and emotional suffer-ing – which computers cannot do.

There's a curious contradiction here that nobody ever talks about: at the same time as science proclaims its moral neutral-ity, proponents of AI want machines to become moral agents. Never more so than with what *Nature* has taken to calling 'ethi-cal robots'.

Ethical robots it seems will come as standard fittings on the driverless cars being developed by Apple, Google and Daimler. They will answer the big questions, automatically …

Should driverless cars be programmed to mount the pavement to avoid a head-on collision? Should they swerve to hit one person in order to avoid hitting two? Two instead of four? Four instead of a lorry full of hazardous chemicals? This is what the 'ethical robot' fitted into each driverless car will decide. How will it decide? In July 2015, *Nature* published an article, 'The Robot's Dilemma', which explained how computer scientists:

> have written a logic program that can successfully make a decision … which takes into account whether the harm caused is the intended result of the action or simply necessary to it.

Is the phrase 'simply necessary' chilling enough for you?

One of the computer scientists behind this logic program argues that human ethical choices are made in a similar way: 'Logic', he says, 'is how we … come up with our ethical choices.'

But this can scarcely be true. For good or ill, ethical choices often fly in the face of logic. They may come from gut instinct, natural cussedness, a desire to show off, a vague inkling, a shudder, a sense of unease, or a sudden imaginative insight.

I am marching through North Carolina with the Union Army, utterly convinced that only military victory over the Confederacy will abolish the hateful institution of slavery. But I no sooner see the face of the enemy – a scrawny, shoeless seventeen-year old farm boy – than I throw away my gun and run sobbing from the battlefield. This is an ethical decision resulting in decisive action, only it isn't made in cold blood, and it goes against the logic of my position.

Computer scientists writing the logic program for an ethical robot may appear as modern as modern can be, but their arguments come from the 1700s. The idea that ethics are logical appeals to what – in another context – Hilary Putnam describes as:

> the comfortable eighteenth century assumption that all intelligent and well-informed people who mastered the art of thinking about human actions and problems impartially would feel the appropriate 'sentiments' of approval and disapproval in the same circumstances unless there was something wrong with their personal constitution.[*]

The thinking may be strictly 1700s, but the technology isn't. The US Department of Defense is at work on tiny rotorcrafts known as FLACs (Fast Lightweight Autonomous Crafts) that will that will be able to go inside flats and houses, office blocks and restaurants and deliver a one-gram explosive charge to puncture the cranium. These FLACs are types of Lethal Automative Weapons Systems. (LAWS). If drones weren't bad enough, LAWS are on a whole new level. With drones, a human always makes the decision whether to kill, from however far away. But LAWS are a break with tradition. They are fully autonomous.

According to Stuart Russell, professor of computer science at University of California at Berkeley, this means allowing machines to choose whom to kill – for example, they might be

---

[*] Hilary Putnam, *The Collapse of the Fact/Value Dichotomy and Other Essays*, 2002.

tasked to eliminate anyone exhibiting 'threatening behaviour'... 'one can expect platforms deployed in the millions, the agility and lethality of which will leave humans utterly defenceless. This is not a desirable future."[*]

Presumably, LAWS just go through the kill list on the drop-down menu until their batteries run out.

If the US Defence Department want LAWS to be a 'success-ful' weapon system, their ethical data entry had better exclude the International Covenant on Civil and Political Rights (1966) with its provisions on the security of person, procedural fairness and rights of the accused. Or else the drone might turn tail and direct its fire at those who gave wings to its eternal mission of unlimited extra-judicial killing.

Delegating ethics to robots is unethical not just because what robots do isn't ethics but binary code, but because no logic program could ever predict the incalculable contingencies, and shifting subtleties and complexities entailed in even the simplest case to be put before a judge and jury. By its very nature justice cannot not be impersonal and still be just. 'Use every man after his desert', Hamlet snaps at Polonius, 'and who shall 'scape whipping?'

Uploadable You

The fact that ethical choices may be prompted as much by a twisting of the guts or a trembling in the arms, may be what leads some to wish to escape our bodies altogether by means of a parascientific vision called 'uploading consciousness'.

---

* Stuart Russell, 'Take a stand on AI weapons', *Nature*, 2015.

'With powerful enough computers simulating the interactions in our brains,' argues David Eagleman in *The Brain: The Story of You*, 'we could upload. We could exist digitally by running ourselves as a simulation, escaping the biological wetware from which we've arisen, becoming non-biological beings.'

Uploading consciousness sounds futuristic but it speaks to those parts of Christianity heavily influenced by Plato and Pythagoras. Beyond bodily sin and earthly decay lies a shimmering non-physical realm to which the enlightened may be uplifted. But to the girdle do the Gods inherit, but from the neck does the software upload. The clergy have been trying to make us non-biological beings for thousands of years. It hasn't worked because their Platonic dream of a disembodied intellect is biologically impossible. We should know this by now. Unlike the ancient Greeks who were forbidden to do autopsies, we have been dissecting human cadavers for millennia, long enough to know that there is no brain without a body.

Even if I am paralyzed from the neck down, I still have complex biochemical feedbacks looping between head and heart, hormones and hypothalamus, guts and glands. Even in my tetraplegic state, I am not just a head on a pillow. I am a body on a bed, and the cure for my bedsores is not beheading.

Hormones from the endocrine system reshape our neural pathways. A good half of the endocrine system is found below the neck in the thymus, adrenal glands, pancreas, testes or ovaries. This makes it incoherent to talk about the brain escaping the body. You can't have one without the other.

In the nineteenth century the great biologist Thomas Henry Huxley reflected how many a beautiful hypothesis is slayed by an ugly fact. Those were the days. What has been happening lately is the strange phenomenon of the beautiful hypothesis enjoying more popularity *after* being slayed by the ugly fact than before.

The discovery that bodily hormones shape neural pathways in teenage brains should have left fantasies about simulating the human brain as obsolete as other monuments to old-time sci-fi such as derelict monorails, and yet these fantasies persist.

Claims that humans can transcend biology proceed as if the last few years had seen not the confirmation but the falsification of neuroscientific theories about, for example, endocrinal influence on the brain, of complex environments on synaptogenesis, of the effects of maternal enrichment of foetal brains. Any or all of these ugly facts (in the Huxleyan sense) should have put paid to ascetic fantasies of uploading consciousness and 'escaping the biological wetware from which we have arisen'. That they did not, tells us something. To steal a phrase from the great geochemist Vladimir Vernadsky (1863–1945) they tell us that uploading consciousness is a 'political idea not a scientific one'.

Or possibly a religious one, as in Ray Kurzweil's *The Singularity Is Near: When Humans Transcend Biology*. The Singularity, says Kurzweil, is when non-biological intelligence takes over. People will transcend base flesh and discover that they can assume an utterly new 'physical manifestation at will'.

The Singularity strikingly mirrors what fundamentalist Christians call The Rapture. Seven years before the Second Coming of Jesus Christ, true believers will be transfigured from their base biological selves into pure spiritual beings, ascending into the air. While unbelievers suffer the 'time of tribulation', the Rapture lifts true followers of Christ away from the carnality of the flesh and mortality.

The Singularity also shares with the Rapture the troublesome necessity of always pushing back the date of when it is supposed to happen. The Rapture's currently predicted date is 2019. (An earlier, much trumpeted ETA of 2011 didn't come off as hoped.) Kurzweil originally slated 2029 as the year of the Singularity, but has since revised the date to 2045. ('No matter how much they dig up the pavement, Ray, the broadband don't get no faster, innit?')

Another reason that AI fantasies are able to withstand so many of T. H. Huxley's 'ugly facts' is because those ugly facts come from outside physics and are therefore beneath attention. Ernest Rutherford's boorish remark that 'there is physics and there is stamp-collecting' has an unstated corollary, which is that the evidential standards of other disciplines are not worth bothering about. This being so, physicists need not trouble to find evidence to support any claims they feel like making about anything whatsoever outside the *sanctum sanctorum* of physics, such as biology, ethology or philosophy.

When Stephen Hawking says 'philosophy has not kept up with science' he doesn't provide any evidence, but then he probably feels he doesn't need to. It's not as if he is talking about anything important like the different grades of space gravel to be found on different planets.

To support his claim, Hawking might cite a drop in publications dealing with the philosophy of science. Or he might produce a graph showing a decline in philosophers and scientists submitting jointly-authored papers to peer-review journals. Evidence that philosophy departments are dropping science modules would be nice. Any or all of these would give some intellectual rigour to his allegation, but why bother? No-one expects any kind of rigour when it's only stamp-collecting we're talking about.

The great irony of scientism, however, is that even as its proponents declare that philosophy has nothing to teach science, their every word is saturated with philosophical assumptions, often hoary ones. That's why they stay true to metaphors based on ideas of nature now known to be false, such as the idea that organisms are just wet machines which we will look at in a moment, but first I want to examine its equally fallacious corollary that algorithms can do everything that lifeguards do.

What do lifeguards do all day?

In *Homo Deus*, Yuval Noah Harari imagines a future in which 'algorithms push humans out of the job market'. It's not just lawyers that are going to be replaced by robots, it seems chefs, waiters, security guards and lifeguards are on the way out, too.

Harari cites a couple of academic papers that give statistical predictions about the future of employment. There will be an 84 per cent drop in the number of security guards, apparently. Automation will also result in a 74 per cent decline in lifeguards.

But what does Harari imagine lifeguards do all day? Dive into the water and fetch people out?

The lifeguards at my local swimming pond are diplomats, care assistants, swimming instructors, cleaners, caretakers, venue managers, goodwill ambassadors and bouncers. Each and every one of these functions is vital to the smooth running of the pond. Not a single one of them could be done by a robot.

Their diplomacy, for example, takes many forms. Sometimes it involves finding a polite but firm way to tell a man that he is too drunk to swim. Tact is also required when telling parents that their child is too inexperienced a swimmer to swim in deep water on her own. As likely as not the parents will blow their top. Their whole plan for the day has been ruined. On top of which they are being told that they do not know their own child's swimming ability.

As care-assistants the lifeguards gently guide the eighty-seven-year-old regular to the water, holding her hand as they walk her along the slippery promontory to the pond's edge, providing a much-needed human touch to her day. They also help a one-legged man retain his dignity by helping him back up the steps after his swim, but so unobtrusively that it looks like he got out of the water without any assistance at all.

My reason for going into such detail about a lifeguard's duties is to highlight the way in which arguments that robots will replace people in the workplace are based on incredibly simplistic fantasies about what people do all day. Rather than hands-on experience of what the job entails, these fantasies have the downsizer's high-handedness. The lifeguard pulls people from the water. A robot could do that. A lawyer recites precedent to a judge. A robot could do that. An AI enthusiast looks at a complex human picture but is only able to assimilate a meagre 3 per cent of what is happening. Now there's something a robot really could do!

Of course, there may well be a 74 per cent cut in lifeguards, but this will have nothing whatsoever to do with drones called Kingfishers that dive into the water and snatch out swimmers who have either got into difficulty or who, like me, have a flailing swimming style that makes it look as if they have. The 74 per cent drop in lifeguards will not be because of automation, and only indirectly because of algorithms. Councils are cutting back on public services to offset the catastrophic losses entailed by the mighty algorithms of a collapsed financial system.

Slaves to the algorithm

There is a touching naivety to so many of the fantasy scenarios of futurologists. In *Homo Deus,* for example, Harari sketches a futuristic dystopia that almost doesn't bear thinking about:

> wealth might become concentrated in the hands of [a] tiny elite… creating unprecedented social inequality.

Well, I shan't live to see it at least!

When you read Elon Musk, Ray Kurzweil or Yuval Noah Harari's dire predictions that algorithms might one day decide to do things according to their own logic rather than human needs and wishes, you want to say, 'where have you been all your life? Aren't you futurologists supposed to be unusually plugged into the *zeitgeist*? How come you haven't seen what everyone else cannot avoid? You seem not to have been allowed out of the house very much. Let me tell you about something called the stock exchange and the bond market.'

'What is good for a shareholder in a firm, you see, tends to be bad for the people who actually work there. Staff may find themselves doing unpaid overtime to fulfil a shareholder's ambition to own one of those automatic up-and-over garage doors, where you can just remotely open it while still driving towards it … like you're in the future already. What is good for people – full employment, say – is bad for bond holders, because full employment leads to inflation, and inflation lowers the value of bonds.'

One more reason for the rich to fantasise that algorithms may push humans out of the job market, I suppose.

Talking of which, the other thing that futurologists haven't noticed but which everyone else has is the exponential growth in the numbers of security guards. They were supposed to be on the way out, remember, but inequality has been a boon for the private security industry. The only way there could be the 84 per cent decline in security guards – as Harari's book predicts – would be an 84 per cent increase in equality. Any takers?

The Last Cyborg

The machine model of the organism was wound up and set walking in the 1600s by Descartes. In the following century, Julien Offray de La Mettrie was very impressed by The Digesting Duck, an automaton with over four hundred moving parts. Invented by an engineer called Vaucanson, the mechanical duck's party piece was that a few minutes after 'swallowing' a piece of bread in its metal bill a green turd plopped out of its metal arse. That green turd was it for Julien de la Mettrie. He'd seen enough.

'Let us then conclude boldly that man is a machine,' he wrote in *L'homme machine*. The Digesting Duck clearly showed that 'to make a talking man, a mechanism no longer to be regarded as impossible,' was simply a matter of more moving parts.

This sort of talk remained more or less scientifically respectable until 1859, when Charles Darwin took a big sledgehammer and smashed the machine model into springs, cogs and sprockets, by way of showing that we are – deep breath – not machines but animals.

Everyone always talks about the hammer blow *Origin* delivered to the biblical story of Creation, which of course it did, but we should remember that it also snatched away science's machine metaphor. This was no small matter of just depriving science of a figure of speech. The machine metaphor was the motherboard of an entire conception of the universe. Snatch it away and a whole world goes with it.

If animals are not machines, if brains do not operate according to the principles of physics, if neither tropical forests nor historical events follow mechanical principles, then how shall we explain their world? It proved too big an ask. After a while Darwin's idea that we are not machines but animals was quietly dropped. Guiltily, like a dog eating its own faeces the Church Scientific went back to its seventeenth-century cyborgs, and has stayed faithful to them ever since, illustrating a much overlooked principle in the history of ideas: when the facts don't fit the story, the facts have to go.

In *The Structure of Scientific Revolutions,* Thomas Kuhn advances his famous thesis that the accumulated weight of discarded evidence, discarded because it doesn't fit the mould, ends up cracking the mould. This cracking of the mould, he says, takes the form of a scientific revolution, which ends up producing a new mould: the famous 'paradigm shift' that made Kuhn's reputation.

Even though Thomas Kuhn's shifting paradigms enjoy widespread acceptance, I'm afraid I don't believe this Hegelian fairytale. I don't believe that history follows organic patterns of growth and development in the way that watercress and slender lorises do. And I do not believe that history follows any kind of predictable rules like a cosmic mechanism. Just because the structure of science has cracked open once or twice before and found a bigger and better shell doesn't mean that science must always act like a growing hermit crab. On the contrary, there is evidence to suggest that, far from cracking the mould, findings that don't fit the old picture simply fall away.

I think the process by which the scientific community selects for and against ideas that fit or don't fit the imaginative vision can be described a little differently from the Kuhnian paradigm shift.

The Glue and Glitter Painting

Rather than paradigm shifts, I think the process is more like primary school glue and glitter paintings. First the child paints a picture of a house in glue on a thick sheet of A3 art card. Four windows and a chimney. When colourful glitter is shaken over the card, glitter that sticks to the shape of the glue becomes part of the picture. Glitter that slides off the paper never becomes part of the house.

If, by the time we are in Year 6, we find ourselves waist-deep in glitter and still producing the same picture we came up with in reception, the famous paradigm shift is supposed to happen. The four windows and a chimney picture is thrown away and all the accumulated glitter on the floor is now used to make a new, more sophisticated, truer, more realistic, finer-grained picture of a house. But sometimes we just scoop up a handful of glitter, and show how it too can be used to make the shape of the four windows and a chimney house. Then we say: '*Look! Doesn't the fitting of the new evidence to the old shape show that the old shape was right all long?*'

The only problem is that it becomes harder and harder to learn anything new in a classroom that is half-buried in multi-coloured glitter.

The Glue and Glitter Painting process can be seen in the rejection of Darwin's idea that we are animals, and the subsequent return to the seventeenth-century machine model. A return which opened the way for the Iron Laws of History that animate Kuhn's *Structure of Scientific Revolutions*.

In neuroscience, the AI-influenced notion that your brain is a computer, of an idea sometimes called 'the computational theory of the mind', is so gorgeously science-y that it cannot be allowed to go, no matter the evidence stacked up against it, and no matter that search parties now have to be sent into the deep drifts of multicoloured glitter to try to locate the sound of buried children tapping on the pipes awaiting rescue.

7. I'M LOOKING THROUGH YOU

Under the headline: 'Neuroscience: I built a brain-decoder', the BBC Future website announced:

> Scientist Jack Gallant can read your mind. Or at least, he can figure out what you're seeing if you're in his machine watching a movie.

The claim that machines can read our inmost thoughts imperils mental health. For this reason, I believe broadcasters and science journals have a duty to be very circumspect in how they report such claims. By lending their authority to headlines such as 'I built a brain-decoder' they leave us believing that our minds have been hacked, that the inner life has been blown open like the walls of a toilet cubicle in a *Carry On* film. To tell someone that you can see into their head is to attack them.

This claim is no less psychologically damaging for its being scientifically false. The press-kit distributed by the brain-decoder's PR team includes a split-screen showreel. Left side of the screen are film clips. Right of screen – we are told – are brain scan images of someone watching that same clip. When screen left shows Steve Martin, screen right shows an eerie Turin Shroud of Steve Martin. When screen left shows a scarlet macaw gliding through the sky, screen right shows a fuzzy red cross gliding over a fuzzy blue background.

'In other words,' says the BBC, 'they took brain activity and turned it into pictures revealing what a person was seeing. The technical process behind how the brain decoder works is explained in a *Nature* article (reprinted in *Scientific American*). The brain scans of people watching a film are:

> fed into a 'pattern classifier', a computer algorithm that learns the patterns associated with each picture or concept. Once the program has seen enough samples, it can start to deduce what the person is looking at or thinking about.

I am grateful to BBC Future for being less coy, here. In fact, it's fair to say that the BBC lets the cat out of the bag good and proper: 'An algorithm then used those signals to construct a fuzzy composite image, drawing on a massive database of YouTube videos.'

Well, now. Well. Now. Well, and indeed, now. Imagine if oceanographers announced that footage of the Mariana Trench was partly filmed by deep-sea submersibles equipped with remotely operated cameras, but also drew on a massive database of YouTube videos. What would be the scientific status of this footage if now and then among the hydrothermal vents we also caught a glimpse of Spongebob Square Pants? How could we be sure those benthic dinoflagellates were not Captain Barnacles?

Could we trust the footage of the angler fish not be Thunderbird 4 with the headlights on? Or imagine a NASA press conference where they confessed that the footage of Mars' surface came from combining pictures from the Curiosity rover's seventeen cameras with a massive database of YouTube videos, such as *The Martian*, *The Man Who Fell To Earth* and *The Clangers*.

If a brain-decoder actually existed, it would be very big news indeed. The fact that it is not big news therefore presents a problem for brain-decoding stories, especially when they get to discussing possible future applications of this exciting new technology that lifts the lid on the brain. The problem is solved by a self-deprecatory coda tempering our giddy hopes for mass-market roll-out of brain-decoders.

Apart from anything else you need a 15-metric tonne, US$3-million fMRI machine and a person willing to lie very still inside it and actively think secret thoughts. Even then, says Gallant, 'just because the information is in someone's head doesn't mean it's accurate.'

No, you don't, Jack Gallant. No, you don't. See what he did? To make this a question of whether what's in my head is accurate or not, takes it as read that he can see what's in my head in the first place. It assumes he can read my thoughts as simply as an ocean explorer looks through a glass-bottomed boat. A less disingenuous caveat would be to say, 'just because the information is on YouTube doesn't mean that it's accurate'. And what stops roll-out of this technology has less to do with its unwieldy tonnage and prohibitive expense, perhaps, than with getting copyright clearance for using clips from old Steve Martin films.

* * *

The Unexpected Visitor

Is everything we know about mammalian vision wrong, or have claims about mental transparency been exaggerated? I ask this because if Jack Gallant's brain-decoder really works in the way that the science journals are happy to suggest it does, then fifty years of science into how human vision works has just been falsified.

In the 1960s, the great Russian psychologist Alfred Yarbus published *Eye Movements and Vision*. In a series of famous experiments he asked people wearing a terrifyingly clunky proto-contact lens – basically a rubber suction cup with a pinhole in it – to look at an oil painting called *The Unexpected Visitor* by the Ukrainian realist Ilya Repin, (who also painted remarkably frank and fresh portraits of Tolstoy and Turgenev). By bouncing light off the pinhole, Yarbus was able to track exactly where the eyes went as they looked at the painting. (Me personally, I would have been casting nervous glances at Yarbus to check if he was dancing about like Rumplestiltskin.)

The Unexpected Visitor depicts a man returning to his family from political exile in Siberia. His family don't seem to know how to react on his return, and he doesn't seem know how to react to them. The power of Repin's picture comes from the way that the unexpected visitor is both strange and familiar at the same time. Yarbus discovered that the way we look at the world is both strange and familiar too.

By plotting a scan path of just what we look at, in what order and for how long, Yarbus revealed that human vision is characterised by saccades, rapidly flickering, unconscious eye movements that dart between fixation points, returning over and over again to points of significance in the picture. The scan path is like the zigzagging scrawl you get when sky-writing with sparklers, or like the airline brochure page where they superimpose all their flight paths onto one map: much to-ing and fro-ing between fixed points, and bumpy around the major conurbations. Our eyes move back and forth over the picture because we are searching for the details that have the most explanatory force, those that will help us understand most fully what is going on. We are searching for the significance in what we see, we are linking things and going back to things until we find a pattern that makes sense of them all. And all in the blink of an eye.

Perhaps a better analogy than either sparklers or flight paths, might be to say that the jerky saccades of the human eye are like the head movements of a bird. A bird's eyes are not as mobile as ours, so the bird tilts its head this way and that as it follows its scan path across a painting linking the parts of a picture into a meaningful whole. Painting?

Painting. Recent experiments in avian perception have discovered that pigeons can unerringly tell a Monet from a Picasso. This goes way beyond being able to recognise a dozen paintings in order to get a food pellet when they tap their beak on the Monet button upon being shown *Water Lilies and Japanese Bridge*. Amazingly, the pigeons can be shown *any* painting from *any* period in the entire *oeuvre* of either artist and still tap their beak on the correct button. Never mind the subject matter, the medium, never mind if it's abstract or figurative, what the pigeon can somehow tell is the indescribable thing which every

Monet painting has in common with every other, and what every Picasso painting has in common with every other, despite the bewildering diversity of Picasso's art. We might call it style, feel, character or *gestalt*, but whatever we call this thing, this is the thing that the pigeon attends to when successfully discriminating between the work of the two painters.

Something similar to the Monet–Picasso pigeon experiments, was discovered by Wolfgang Köhler's famous 1940s experiments with chickens and shades of grey. Köhler showed that vertebrates have evolved to see the relations between objects in a field of vision, and in fact see things only in the context of other things. (Dick Swaab says this is something Americans cannot do, you will recall, but doesn't specify whether this inability extends to American chickens, too).

Köhler placed chicken feed on the lighter of two squares of grey paper. He did this every day until the chickens no longer bothered inspecting the dark square. He then replaced the dark square with a square that was an even brighter shade of grey than the one upon which the pigeons were accustomed to finding their chicken feed. Next morning the chickens went straight to the unfamiliar bright new square. But the chickens were doing nothing new. They were repeating what they always did, simply going to the brighter of the two. Köhler concluded that birds see objects in terms of their relation to each other. There are sound evolutionary reasons for having 'brains attuned to gradients rather than to individual stimuli', as E. H. Gombrich points out:

> For would a memory of the exact stimulus have helped them
> to recognise the identical paper? Hardly ever! A cloud passing

over the sun would change its brightness, and so might even a tilt of the head, or an approach from a different angle. *

But Jack Gallant reverses Wolfgang Köhler's finding. Our brains are not attuned to gradients but to individual stimuli after all. And where Alfred Yarbus's experiments demonstrate that we are active searchers after meaning, Gallant's brain-decoder reverse that as well. Rather than saccades darting between recurrent fixation points, Jack Gallant shows that human vision cuts between close-up, dolly zoom, fixed camera tracking shot, slow pan and Steadicam single. (All of which suggests that the brain-decoder represents not what you see but what you are shown.) We are not active meaning-makers, but passive blank screens. Films projected onto the inside of our skulls will come back a little fuzzy but otherwise just the same, a bit like if you dropped them in a puddle.

You'd think the science journals would devote whole issues to the breaking news of how the brain-decoder has led to the greatest revolution in our understanding of vision since Isaac Newton's *Optics*. You'd think they'd report how universities worldwide are rewriting textbooks in the light of the brain-decoder's revolutionary discovery that the eye works according to the camera styles of whichever film or TV genre we happen to be watching.

I guess when so many scientific revolutions are happening all at once there's no time to stop and count the fallen theories along the way. For in the same article in which BBC Future and

---

* E. H. Gombrich, *Art and Illusion*, 2002.

Nature break the brain-decoder story, they also drop the bombshell Yukiyasi Kamitani has developed a dream-decoder too.

Kamitani claims that his dream decoder can itemise two out of every three objects in our dreams. Kamitani records patterns of brain activity when you look at, say, a horseshoe, and then if the same pattern appears while you are sleeping, then – hey presto! – he can tell that you are dreaming of a horseshoe. By cross-referencing signature patterns of conscious brain activity with those recorded while we are sleeping, Kamitani knows what we are dreaming about.

Except …

In dreams, objects tend not to be reliably themselves. The people, places, objects in our dreams neither look nor behave like they do in the real world. An eight-legged goat has the voice of a much-loved primary school teacher. She is very cross with you for hiding top-secret documents in a village fete tombola. 'Don't you know there's a war on?' she asks. An egg ticks like a clock. Bono is extremely worried that the Irish Treasury may not receive enough tax revenue to provide essential social services. Nothing is what it seems!

Houdini versus the Brain-decoder

Houdini (1874–1926) the world's greatest ever escapologist, devoted his retirement to helping the rest of us escape chains of illusion. To this end, he made it his business to expose mind-readers, psychic mediums, telekinesis and spiritualists. I would have loved to let Houdini loose on the brain-decoder and dream decoder.

Having shared vaudeville bills with the great mind-readers and illusionists, Houdini knew that there was no such thing as mind-reading. Besides, if minds could be read then these brilliant acts wouldn't have been half so brilliant as they were. Among Houdini's peers were mind-reading double-act Handy-Bandy and Nadya Nadyr, and mind-reading solo performer Newmann The Great ('Ask Him He Knows').

The clairvoyant Newmann (no relationn) answered questions from the audience. Newmann's posters were studded with suggestions of what the audience might want to ask him: 'Is My Sweet Heart True?'; 'Where Is My Watch?'; 'Will I Win My Lawsuit?'; 'Who Stole My Hat?'; 'Is May My Enemy?'. That last one makes you think Newmann really could see into the future after all.

Houdini and his peers knew not to make the illusion look too easy. If, when Handy-Bandy holds up a picture of King Kong climbing the Empire State Building, a blindfolded Nadya Nadyr draws a faithful reproduction of the famous film poster the audience will detect a swizz. But if Nadya Nadyr draws a monkey climbing a tree the audience will gasp. They will gasp because they sense that something has indeed been sent from Handy-Bandy's mind to hers, only the signal was indistinct, the telepathy wavelengths impure, a disturbance in the ether somewhere between Handy-Bandy's forehead and Nadya Nadyr's sapphire bindi. The great thing about getting it ever-so-slightly off is that you recruit the audience into making the trick work for you. They fill in the gaps, and by so doing leave the jury box for the witness stand, so to speak, testifying on your behalf. This trick of mind-readers is a kind of *sfumato*, the smoky or blurry effect of artfully smudged paint. The monkey up a tree is a *sfumato* version of King Kong up the Empire State Building.

The impressionistic right screen outline of a man who may or may not be Steve Martin is compelling precisely because it is so blurry. It is the monkey climbing the tree rather than King Kong up the Empire State Building. The proof of telepathy. But if the right-hand screen shows the star of *The Man With Two Brains* too clearly, it reveals that the only brain we are watching at work is Jack Gallant's.

To be clear, I am not in any way saying that Jack Gallant has deliberately degraded the image. What I am saying is that part of the allure that Gallant's images have for us, comes from the *sfumato* which enlists our help in joining the dots. This is not a trick of deception. It is in fact no different from the way that fMRI gets our attention by using the semaphore colours orange and yellow to indicate the fusiform gyrus 'lighting up', even though it doesn't light up at all, let alone in orange and yellow. No-one ever planned for fMRI to have those semiotics. Back in the early days of fMRI, no-one ever said: 'Let's use the same colours for voxels as they use on semaphore flags. That way we can make people feel that the brain is trying to send signals to them.'

Nor did Jack Gallant deliberately fuzz things up like the production on The Jesus and Mary Chain's *Psychocandy*. He never said: 'Hmm, this will look more convincing if I make it more opaque, or just drop it in a puddle.'

I am not accusing Jack Gallant of charlatanism, I'm saying we should catch onto ourselves and think about why we are so tempted by this psychocandy. I am also deploring the way that national broadcasters and science journals think nothing about telling us that private life is over, when the evidence seems to be wanting, not least because it has recently become far from clear that facial recognition happens in the part of the brain that we thought it did. Brain and dream decoders assume the existence of a dedicated face recognition area but that is proving to be a slippery concept

Moonlighting in the fusiform gyrus

Within the long thin slither of the fusiform gyrus, the fusiform face area (FAA) was once supposed to be the bit of the brain that recognises faces, and detects shifts in facial expression, changes of mood and so on. Facial recognition has since been demonstrated to be an activity widely spread throughout the brain. The FFA has lost its monopoly on the facial recognition business. Perhaps this is why it has responded by diversifying. For the fusiform face area has been discovered to be involved in more than recognising faces.* Instead of it being solely a face recognition area, it's more accurate to say that it's where fine distinctions are made. This is where Roberta Vinci spots the difference between Coco Vandeweghe's disguised backhand topspin and her disguised backhand slice. This is where apple experts distinguish a Belle de Boskoop from a Worcester Pearmain. The fact that the FFA has been moonlighting, doesn't mean that it has given up on its core business. It is still a face recognition zone, but it also helps out wherever nice distinctions are to be made.

---

* Tarrand Gauthier, 'FFA: a flexible fusiform area for subordinate-level visual processing automatized by expertise', *Nature Neuroscience*, 2000.

The lack of a hard and fast correspondence between the fusi-form gyrus and facial recognition should raised doubts about the ability of machines to know when we are looking at faces. The original claim, if you recall, was that the brain-decoder records our brain activity when we look at faces. Then in a blind trial, Jack Gallant's team need only observe a matching pattern of brain activity to know we are looking at our face.

But how does this work when facial recognition takes place in brain areas every bit as varied as the quality of Steve Martin films?

The Lying Lie-Detectors

I don't know why belief in the existence of lie-detectors or poly-graphs is any less embarrassing than a belief in flying saucers or telekinesis, but for some reason it seems to be more socially acceptable. The single fact that lie-detectors do not provide evidence acceptable in any court of law anywhere in the world proves beyond a reasonable doubt that they haven't actually been invented yet. Lie-detectors can't detect lies, but they do have others. Firms use them to make their staff feel small, in what psychologists call a 'degradation ceremony'.

Institutions use degradation ceremonies to induct raw recruits into a proper understanding of the hierarchy. Prisoners have their names taken away from them, job centres keep you waiting, and doctors ask you to drop your trousers. Showing is stronger than telling. In this way, those with power simply show the inductee their powerlessness. The firm uses a polygraph not to read your mind – which the machine cannot do – but to control you mind. The machine cannot do that either, of course, but the degradation ceremony can. Management wants you to think they know you inside out.

A firm called Sheffield Detectives offers companies a pre-employment polygraph test to assist in the hiring of new recruits. Their 'comprehensive pre-employment security interview' goes into everything from a prospective employee's education to their 'morality and illegal drug use'. In an Orwellian twist, Sheffield Detectives' brochure says that 'trust is an important factor in any business'.

At the same time as companies are advised not to trust their employees an inch, they must be very trusting toward the claims advanced for a truth-reading machine. This bizarre combination of naive trust in technology and extreme suspicion of outsiders is found not just in Sheffield, Yorkshire but also in Langley, Virginia, home of the CIA: 'We discovered that there were some Eastern Europeans who could defeat the polygraph at any time,' said Richard Helms, former US Director of Central Intelligence. 'Americans are not very good at it, because we are raised to tell the truth and when we lie it easy to tell we are lying.' The head of intelligence under Richard Nixon continues: 'But we find a lot of Europeans and Asiatics can handle that polygraph without a blip, and you know they are lying and you have evidence that they are lying.'

If evidence exists to prove that the European or Asiatic suspect is lying, then who needs a polygraph? What possible function can the polygraph serve that evidence does not serve much better? Why does evidence from the outside world seem insubstantial and unreliable compared to the polygraph? At this point, it might be useful to have a look at how exactly a polygraph works.

Polygraphs don't measure actual lies, they measure sweaty palms and heart beats. Now the term sweaty palms is, as you might imagine, eschewed by polygraphists. Anxious to put a bit of clear blue water between themselves and witches or homeopaths, there is a solemnly observed convention among psychologists and neuroscientists always to prefer the phrase 'galvanic skin response' (GSR). Much more scientific. Authoritative. What lie-detectors measure are galvanic skin responses (GSRs) and not sweaty palms (SPs).

The uselessness of polygraphs to the criminal justice system, however, can be seen by listing some of the things that can trigger a positive GSR. The list includes: fear, stress, confusion, nerves, mental illness, drugs and drug withdrawal, alcohol and alcohol withdrawal. I ask you, is it possible to imagine yourself sitting in a police cell without ticking at least one item in the list? In fact, only the habitual, career criminal would feel no nerves, no fear, and no confusion. Only the ice-cold Mr Big knows exactly why he's been arrested. But it seems the rest of us can only fail the lie-detector test. The degradation ceremony has worked it magic. They have got us where they want us, ready to sing like a canary.

Many news articles are very excited by lie detection in the age of sophisticated brain-imaging. Lightweight portable fMRIs will soon be able to detect those regions of the brain that light up when felons lie. The game is up for crooks and liars. Except there is no one region of the brain devoted to lying. In *Brainwashed: The Seductive Appeal of Mindless Neuroscience*, Sally Satel and Scott O. Lilienfeld beautifully illustrate this by listing every brain area that has, in one fMRI test or other, been implicated in lying or deception.

> [T]he array of brain regions correlated with deception is dizzying: the parahippocampal gyrus, the anterior cingulate, the left posterior cingulate, the temporal and subcortical caudates, the right precuneous, the left cerebellum, the anterior insula, the putamen, the thalamus, and the prefrontal regions (anterior, ventromedial, and dorsolateral), as well as regions of the temporal cortex.

That's a lot of brain! It makes you wonder whether all these flashing lights in the brain might be connected at a deeper level. Perhaps when we zoom in another few microns they will turn out to be the fairy lights of a Mississippi paddle steamer of the mind. Leaning against the stern rail, a riverboat gambler idly flips the fairy lights on and off with his pearl-tipped walking cane. Sensing our macroscopic gaze, he looks up at the full moon of our electron lens.

'So you wanna know if I am the source of all the lies in your brain', he drawls. 'Let's say we play a hand of poker to find out? Question you need to ask yourself is … will I know what cards you're holding, me being your brain and all? You desire to know what the stakes will be. Well, now, if I win, let's say we put some of your emotional baggage ashore, huh? Such as still beating yourself up over not marrying Flavia Korova. Like that would have worked. Time to stop lying to yourself, boy. Whaddyasay?'

* * *

The seductive promise that brain-reading might supplant the need for evidence in the outside world has to do with the general idea that humans are hopelessly flawed in a way that computers are not. Brain books and articles riff endlessly on the fallibility of human memory, on how eyewitness statements are always cock-eyed. The Human Disparagement Industry makes us feel that humans cannot plan a society rationally. Better to delegate this onerous task to computer algorithms. We'd only make a mess of it. This line of thinking fits in very nicely with faith the wisdom of the markets. A faith that never yet wanted for a platform.

8. PAVLOV'S DISOBEDIENT DOGS

During the World War II, the Germans blockaded Leningrad for two and a half years in an attempt to starve the population into submission. The 880-day Siege of Leningrad forced terrible privations and suffering on people. First they ate all the zoo animals, then all the stray dogs, and then all the pet dogs. By the end of 1942, however, those good times were over. Things grew desperate. To make the bread ration of 125 grams (4.4 ounces) per person per day go further, the besieged citizens mixed it with sawdust. Soon people were making soup from shoe leather, boiled grass, acorns and twigs. And there were rumours of the cannibalisation of the recently deceased.

One day, during the cold hard January of 1943, Rita Kitanova, a researcher from the Institute of Experimental Medicine revealed to a select group of friends that she had been holding out on them. She had the keys to a secret store of livestock: Pavlov's dogs.

At first her friends didn't believe her. If there were any dogs still alive in this city, they would have heard them bark. No-one had heard a dog bark in Leningrad for over a year. If Pavlov's dogs were still alive, then why had no-one ever heard them bark. Pavlov's dogs were never heard to bark, Rita Kitanova replied, because they were kept inside the Tower of Silence, the sound-proofed, three-story research facility designed by Pavlov to prevent outside noises contaminating his experimental results. 'The windows were covered with extra-thick glass and the rooms had double-steel doors that formed an airtight seal when closed." Even inside the Institute building itself, Rita Kitanova explained, you still couldn't hear the dogs bark, not until you were actually within the inner sanctum of the Tower of Silence itself.

Her small circle of friends were overjoyed. For the first time in over a year they were going to eat fresh meat. They decided to enjoy this rare meal in style. With frying pans hidden under thick winter coats they crunched over sludge and snow to Pavlov's laboratory on the banks of the River Neva. In their pockets were secreted whatever condiments they had managed to find. Rita Kitanova had a dried bulb of garlic in her glove, a friend's waist-coat pocket hid a withered and crumbly parsley root, another friend had pepper in her hat band and linen napkins stolen from the derelict kitchen of the famous Palkin restaurant.

When they entered the Tower of Silence, the smell of live animal and thoughts of freshly cooked meat were overpowering. They began drooling at the mouth. The dogs gave them a look as if to say: 'You're supposed to wait for the bell'.

Ivan Pavlov didn't work with just any dog. A very long process was required before he selected Avgust, Pingel, Postrel, Toy, Bes, Milord, Max, Vampire and Umnitsa. Most dogs didn't

---

* Duane and Sydney Schultz, *A History of Modern Psychology*, 2012.

want to know. They resisted programming in all sorts of ways. When some dogs had the rubber tube slotted into the surgical incision cut into their cheek, they became so dry-mouthed with fear that Pavlov's experiment was the one time in their whole lives that they never salivated at the thought of food. Some dogs would only work with one particular researcher, but no-one else, and when that favourite researcher left, they became listless and uncooperative. Sensory deprivation made many dogs too depressed to work. No runaround, no contact sniffing of other dogs, in fact precious little in the way of smell at all. For a dog having nothing to smell is like being kept in a dark room, but Pavlov was keen to rule out any possibility that a stray smell might elicit pre-emptive salivation in the dogs before the dinner bell was rung. To this end he concocted a sterile odourless environment. Alas, the dogs had not read their Eagleman and so weren't to know that they invented smells within their own brains, and so an odourless cell was no deprivation whatsoever. These were less enlightened times, and many dogs simply grew too unhappy to work with.

A very long selection process was needed before a small sub-set of dogs could be found to fit Pavlov's experimental design. Given such a long selection process a conscientious scientist might wonder whether what was being measured had anything to do with what is found in the real world. After all, when Sir Francis Bacon advanced the empirical method, did he have in mind the construction of an artificial set-up designed to yield pleasingly quantifiable results?

Now, retreating to a Tower of Silence to study so boisterous an animal as a dog may seem self-defeating, rather like retreating to an ivory tower to study elephant conservation, but Pavlov designed the three-story building with the fairy-tale name after the outside world came crashing in one day and almost destroyed his life's work. On 23 September 1924, heavy rains caused the River Neva to burst its banks and flood Pavlov's Institute of Experimental Medicine, located right on the river bank. Pavlov's biographer Daniel Todes vividly describes how 'the rescuers found the animals' cages filling rapidly with water. Pressed upward toward the wire ceilings, the dogs strained to keep their noses above water'.

The cage doors were underwater. Each rescuer grabbed a dog, and dived down into the dirty floodwater, swam through the submerged door to surface in the corridor, they and the dogs gasping for air. Unlike the rescuers however, the terrified dogs didn't know why their heads were being dragged under the water, and fought against the keepers who they believed were trying to drown them.

Every last dog was saved, but many were never the same again. The flood left in its wake, not just tide marks on the wall, but an ornery dog pack no longer willing to play the game. Many dogs – including star pupil Avgust – no longer salivated when the bell was rung.

Had the flood simply wiped their conditional reflexes in the same way that a shock snaps someone out of hypnosis? Or was the cause of Avgust's newly acquired disobedience his loss of trust in his handlers? Pavlov wasn't sure either way. All he knew was that after the flood Avgust and other dogs reacted differently to the bell. Instead of salivating, Avgust now thrashed his head this way and that, trying to escape the wooden stand to

which he was tied. Other times the bell sent him into a catatonic trance.

Before the flood, Ivan Pavlov considered the great significance of his world famous dog experiments to be the creation
of an artificial link between the 'cortical and the sub-cortical',
korkovyi i podkorkovyi. He had hacked through the impenetrable jungle that divides conscious from subconscious, psychology
from physiology, mind from matter. *

Pavlov's fame rests on the myth that he tapped the dogs'
subconscious like you would tap a Malaysian rubber tree. A small
incision and then he waits for the white subcortical sap to drip
into the cup, drop by drop. With these cups of subcortical latex,
Pavlov showed, he believed, that from now on behaviour could
be modified, psychology shaped to fit requirements. Goodbye to
all those vague notions so beloved by previous tender-minded
generations: the innate, the essential, the intrinsic. Now came
the era of psychological reconstruction, rebuilding the brain.
All these ringing declarations, all these magnificent results, all
the bold project of remaking organisms in a planned and scientific way, all this internationally renowned science depended,
however, on banishing the outside world.

The floodwaters, then, broke much more than the lab's river-
side windows. They also burst Pavlov's experimental bubble, and

---

* In Pavlov's defence, I should say that he is not quite the rigid determinist he is painted. The
heavier tones of this portrait, in fact, come from a simple mistranslation. Pavlov's original idea
of conditional – *uslovnyy* – reflex, ie. contingent upon an earlier stage of the reflex arc, was
mistranslated into English as conditioned – *uboslovlennyy* – reflex, ie. determined, hard-wired.
However slovenly this mistranslation, it still marks only a change in emphasis, a change of degree
not kind.

raised a terrible question: were these results only ever applicable in the contrived conditions of the Institute of Experimental Medicine, but not in the real word?

Sure, the results were sound as far as they went. Yes, with a bit of training and a bell, input x will give you output y. But even so limited a result, the flood showed, turned out to depend on a broad array of factors, including bonds of trust and reciprocity built up between dog and handler, and, as we saw earlier, the deselection of unhappy, disobedient or stir-crazy dogs in favour of especially biddable dogs.

When the river washed away every last cup of canine saliva, an open-minded scientist might have glimpsed, among the wreckage, an epistemological challenge. If dogs were machines, could they now be programmed to forget their trauma? If the terror of drowning had wiped one set of reflex circuits clean, would it now be possible to put a new set of reflexes in? Could Avgust learn to associate a softly ticking metronome with feelings of immense assurance? Could an electric buzzer or a flashing light induce the removal of anxiety and fear in a traumatised hound? But Pavlov didn't think like that. He was the sort of man for whom the right response to the outside world bursting your bubble was to build a bigger, more impregnable bubble. The following year, 1925, construction began on the Tower of Silence.

To prevent contamination from the outside world is properly a central concern in all laboratory design, part of a scrupulous attempt to ensure controlled laboratory conditions for scientifically repeatable experimentation. But with Pavlov it goes beyond that. His Tower of Silence, I suggest, has both an ideological and a psychological cause. Ideologically, once you have committed yourself to the doctrine that animals are machines, you will have to build some sort of shed for these faultless contraptions or else they will start behaving like animals. And that's no good to anyone.

Psychologically, we all build Towers of Silence to which we retreat when everything gets a bit too much, but Pavlov was never one to hold with that sort of touchy-feely mumbo-jumbo. Not only did he ban the use of psychological terms in his laboratory, he used to *fine* his researchers for using them.

What are we to make of this? If we accept him at his own estimation, the banning of psychology from the lab was done for pedagogical purposes, to instruct his research assistants in the proper application of Conway Lloyd Morgan's (1952–1936) law for the interpretation of animal behaviour. Morgan's Canon states that:

> In no case may we interpret an action as the outcome of the exercise of a higher psychical faculty, if it can be interpreted as one that stands lower in the psychological scale.[*]

Pavlov insisted that a physiological explanation, unlike a psychological one, was objectively measurable, and for that reason and that reason alone he would forbid any mention of any psychological terms from being used in his hearing. But we need not take him at his word. There may be other reasons why a man who built a Tower of Silence would be irritated on hearing colleagues mutter phrases such as 'sublimation', 'defence mechanism', 'projection', 'transference', 'loony' and 'nutjob'.

---

[*]   C. Lloyd Morgan, *Introduction to Comparative Psychology*, 1894.

Donald O. Hebb gives another example of Pavlov retreating to a metaphorical Tower of Silence: 'Apparently, Pavlov isolated himself from the contemporary literature, and his theory took no account of psychological discussions after 1900.' This sort of behaviour comes at a cost, as Hebb makes clear: '[Pavlov's] theory has not been rejected because it is too physiological, but because it doesn't agree with experiment.'*

You can contrast Pavlov's attempt to insulate his dogs from the outside world with Rosenzweig's account of failing to get a new set of results by introducing indoor lab rats to the outdoor Field Station at Berkeley.

> We tried ... to ask how our enriched laboratory environment might compare with the natural environment ... Groups of a dozen male laboratory rats thrived in the outdoor setting and ... dug burrows, something their ancestors had not been able to do for more than 100 generations ... Unfortunately, however, in the outdoor setting the rats became too savage to handle, so we were unable to conduct behavioural tests with them.

This cheerful admission of defeat seems to me to have more of the authentic ring of science to it than Pavlovian scientism. I know that being bitten on the finger by a feral rat does not compare to having your entire lab flooded, but Rosenzweig's free and open acknowledgement of what he *hasn't been able to show* is made possible by the robustness and significance of what he

---

* D. O. Hebb, *The Organization of Behaviour*, 1949.

has been able to show. If your hypothesis is sound, then even when experiments go wrong they can still offer powerfully corroborative data. In post mortem, the brains of these feral rats showed greater cortical development than any of the indoor lab rats, no matter how complex their environment:

> This indicates that even the enriched laboratory environment is indeed impoverished in comparison with a natural environment.

Pavlov's results were less robust which left him less able to entertain the possibility that the stultifying kennel conditions stunted the dogs' responses to the extent that experimental inference was unreliable.

The bad behaviour of Pavlov's dogs, their refusal to cooperate after the flood, is very good news for us all. Just as the Tower of Silence once did for dogs, so the noisy tunnel of fMRI is supposed to show that we are crude input-output machines. Where Pavlov had rubber measuring cups to catch drops of saliva, brain scans now catch colourful blobs of love, hate, joy, sorrow and other delusions. And so, a century on, we have reason to be immensely cheered by Pavlov's disobedient dogs, Avgust, Pingel, Postrel, Toy, Bes, Milord, Max, Vampire and Umnitsa, which means 'clever fellow.'

9. 'SCIENTISTS DISCOVER THE LOVE SPOT'

At the beginning of 2014, I was one of thirty-five volunteer subjects who took part in a brain-imaging experiment at University College London's Galton Lab. What nobody knew at the time was that this was going to become a famous brain-imaging experiment when it was published as a paper called 'The Neurobiology of Romantic Love', which got international news coverage, typified by the *New York Times* headline: 'Scientists Discover the Love Spot'. And so it stuck in my craw that out of thirty-five volunteer subjects, I alone was written off as a negative result.

So I just want to tell my side of the story.

We were each told to bring along with us to the Galton Lab four photographs: one photo of someone that you're deeply in love with, and three photos of folk you are fond of but not deeply in love with. Then they hook you up to an electroencephalograph (EEG) because they're tracking blood flow to see which brain regions are demanding oxygen from the blood when you look at which photo. If there's a bit of the brain that lights up when and only when you look at the photo of the person you are deeply in love with then – BINGO – they've discovered the bit of the brain responsible for romantic love.

But when you look at photo of someone you're deeply in love with you have all kinds of emotions, including guilt, regret, shame, fear, anxiety, delight, joy. All kinds of emotions and all kinds of thoughts too. I'm looking at this photo and I'm thinking: 'Is this the best picture of me I could have brought?'

To control for what they call 'background neuronal activity' you have to look at each photo twenty times. Now with the best will in the world, after the fourteenth or fifteenth time of looking at the same photo you are not feeling the love. I confess that after a while I was going through the motions. And so I was astonished to hear a 'huzzah' from the control booth. The next moment, three neuroscientists came barrelling out the booth with a great cry of exaltation.

'Oh wow,' they said, 'that's the strongest correlation we've ever had! That's what we call a golden spike! We've never had such a strong reading. Now, just to confirm, which of the photos were you looking at?'

'I'm really sorry,', I replied, 'I wasn't actually looking at any of the photos.'

'Well never mind,', they said, 'which of the photos were you thinking about?'

'I'm very sorry, I was thinking about something else altogether.'

'Yes?'

'If you must know, I was actually thinking about how you've set this experiment up. I was thinking neural activity lasts milliseconds and you're measuring blood flow. That blood has got to go through the heart, I suppose, down to the toes and up and around and won't arrive in the brain until three or four seconds too late. Combine that with the fact that millions of neurons need to fire simultaneously for there to be any detectable change in blood flow, then it's as if you're taking an aerial reconnaissance photograph of Greater Manchester on a Wednesday to find out what was happening backstage at the GMEX the previous Saturday. But don't be discouraged. Don't be downhearted. Clearly your machine has picked up something here. Maybe what we've discovered is the bit of the brain that lights up when we spot an elementary conceptual blunder in experimental design!'

And thus I was written off as a negative result.

I was escorted from the premises by a lab assistant. As she led me down one corridor after another she said not a word. And then it hit me: she thinks I'm a monster. The other thirty-four volunteers were all found to have the full complement of human emotions. I was probed and found wanting. No love spot there. Just gristle, just a gap. I'm the negative result. She thinks I'm a sub-human freak of nature! She thinks I'm a monster!

In an effort to humanise myself in her eyes I commented on a floral display, but the words that came out of my mouth were: 'Pretty flowers!'

Now I even sounded like Frankenstein's monster, or the Hunchback of Notre Dame.

Down one corridor after another she led me. Still not a word. Then, at the end of a long corridor, she cracked a fire exit, turned, fixed me a look and said: 'If you ask me, those fellers are looking for love in the wrong place.' She stuck out a hand and introduced herself as Glynis. I was so grateful to be accepted back into humanity's warm embrace that I took her hand in both of mine, held it to my cheek and said: 'Soft skin of the kind lady!' Then just as suddenly I snatched it back again.

'Wait a second', I cried. 'You work for this laboratory too. So if they are looking for love in the wrong place where does that leave you?'

'No, I'm not part of that project at all', she answered. 'I just have to assist on other team's experiments as a condition of being able to use the lab's facilities for my own research into the neurobiology of guilt. In fact we're always looking for volunteers – would you consider taking part in one of my experiments?'

'No, I wouldn't want to do that.'

'Why not?' she asked.

'Well, to be honest, I don't really think the brain works like that. I don't think there are these different discrete bits of brain that each do different jobs.'

'Oh, that's very interesting,' she said. 'Your view is what in neuroscience is called "wrong". Let's meet up in a week's time and I'm sure I can convince you to change your mind.'

10. THE NEUROBIOLOGY OF GUILT

One week after taking part in the experiment to discover the neurobiology of romantic love, I met up with Glynis in a cafe near UCL.

'Do you still persist in your obstinate refusal', she asked me, 'to accept the fact of cortical localisation?'

'There has been movement on that', I told her.

'Glad to hear it.'

'I do now concede that different bits of the brain do different things – I just don't believe they do it on their own.'

'You're still wrong', she said. 'And you've been wrong since 1848, and the famous case of Phineas Gage.'

As readers may deduce, we fell into a big row about Phineas Gage, in the course of which I put forward my argument about how the conventional tellings of that accident are based on the melodramatic fallacy of the Jekyll and Hyde brain, which is a profoundly un-Darwinian idea.

'And that, Glynis, is why I do not wish to take part in your experiment to find the neural basis of guilt, because I believe that neuroscience, as presently construed, is predicated on a version of evolution that owes but little to Darwin.'

'Oh, really? Un-Darwinian? Me? If that's what you think then it is time for me to show you my treasure.'

She took me round to her flat. From a glass display cabinet she produced a terracotta tree frog that had once belonged to Charles Darwin. She told me he was given it during the voyage of the *Beagle* in Patagonia by indigenous Tierra del Fuegians. Upon his return to England he placed this terracotta tree frog on the corner of his desk in Down House in Kent, where it remained to his dying day. Glynis subsequently won this terracotta tree frog as first prize in a science competition.

I was very touched that she'd wanted to show me her treasure, and as we left the flat I slipped it into my pocket – because I'd had a great idea for a visual gag.

'You know there's a legend', I told her as we walked to the park, 'that those Tierra del Fuegian terracotta tree frogs sometimes come to life.'

We reached a cafe in the park and sat on a terrace underneath a giant hornbeam tree. While we were sitting there having tea and sandwiches, I asked her a question which had been on my mind for quite a while.

'Why do you think people who write about brain science are such misery guts?'

'You seem to be labouring under a complete and utter misapprehension as to what the job of a scientist is', she replied. 'You seem to think a scientist should be like a comedian, and give everybody uplifting, cheery thoughts and send them on their way with a smile on their face.'

'I have *never* believed that to be the job of a comedian.'

'The job of a scientist', she said, 'is unflinchingly to report the facts without fear or favour, and without any regard as to whether the results happen to be edifying or not.'

'I quite agree, but that's not what's happening here, is it? This is a pre-emptive pessimism in search of its own corroboration. This is pessimism as a starting place, pessimism as a manly occupation. It's perfectly respectable for pessimism to be where you end up, but not to be where you start from. A friend of mine is of a naturally sunny disposition but she lives on a very polluted road opposite a petrol station, and is convinced her twin toddlers will both grow up with major respiratory ailments. This always makes me feel very guilty by the way because I grew up in a little village, in a quiet cul-de-sac. Every morning my mum used to put me in the pram, and push me out onto the front garden. At lunchtime, she'd bring me in, feed me, and then push me back out onto the front garden again until it grew dark. My mum always said that this daily regimen of continual fresh air is the reason I enjoyed excellent health as a child – although I didn't actually learn to walk until I was ten years old.'

It was Glynis's turn to get the tea and cakes in. I waited until she'd gone into the cafe building, and then I climbed up the hornbeam tree, propped the terracotta tree frog in one of its branches, then climbed back down again and sat at the table as if nothing had happened.

Just as Glynis was coming out of the cafe – and luckily she was looking down at her tray – a gust of wind blew the terracotta tree frog off its branch. It fell and shattered on the patio. I just had time to kick the smithereens into a storm drain before she returned.

What made it even worse is that over the coming days and weeks she suspected everyone except me. Because I was new on the scene, I was the one in whom she confided her suspicions. I saw her accuse her flatmates of stealing it, then her flatmates' boyfriends. I was feeling just dreadful. I thought I couldn't feel any worse until she started spending all her time trawling through online antique trading sites. So much so, in fact, that her work began to suffer. It got to the point that she was in danger of losing her project's funding because she hadn't set up any experiments.

'Glynis,' I told her, 'I just want you to know that I am now willing to volunteer for your experiment into the neural basis of guilt.'

Visibly moved, her voice breaking with emotion, she replied: 'You've no idea how much this means to me right now. You're the only person I can count on in this difficult time. You are wonderful.'

'Yeah', I replied weakly.

The next day we were back to University College London's Galton Lab, where we first met.

'Okay, pop this electroencephalograph on you head,' she told me, 'while I go into the control booth.' I heard her distorted metallic voice coming through the headset attached to the electroencephalograph. 'Can you hear me all right?'

I spoke into the mouthpiece attached to the EEG: 'Loud and clear.'

'What I want you to do now', she said, 'is to think about something really bad you did quite recently, something which caused someone you care about a lot of grief.'

I heard the EEG hum and whir. I saw her face in the control booth lit up by flashing coloured lights of her instrument panel.

'That's odd', she said. 'Instead of one bit of the brain lighting up, points of light are flashing all over both hemispheres.'

'Well, it's like I've been trying to tell you, there's more connections in the human brain than there are atoms in the universe.'

'Keep thinking about that really bad thing you did to someone not so long ago', said Glynis.

I focused. The machine hummed. I heard her exclaim: 'Wow! It's the same dots of light all over both hemispheres all in exactly the same places. Let me freeze the frame, join these dots with a felt-tip and see if they make any sort of pattern.'

Then there was silence. When she next spoke her voice was at half speed.

'That's very strange ... They appear to be in the shape of a tree frog.'

11. WHY THE LONG FACE?

To repeat the question I asked Glynis: Why are brain science pundits devoted to such a peculiarly pessimistic world view? Why the long face?

I think pessimism as an intellectual fashion among science writers can be blamed in part on Sigmund Freud's *General Introduction To Psychoanalysis*. In this essay, Freud is the first to advance the following argument.

> The three greatest scientific revolutions of all time each delivered a hammer blow to humanity's naive self-love. Heliocentrism takes us from the centre of the universe to a remote speck. Evolution shows that we are made not in the image of God but orang-utang. Freud's own psychoanalysis shows that far from being rational creatures, humans are at all times driven by dark irrational unconscious impulses – apart from him writing that then, of course. But everyone else? *Mental!*

Freud's Theory That Great Scientific Revolutions Are Always Bummers, has never been more influential than it is today. And yet almost everything about Freud's history of scientific ideas is false.

Take heliocentrism. Aside from a handful of professional theologians, did the discovery that the earth goes round the sun ever ruin anybody's day? For a start, pre-Copernicus, nobody except an extreme heretic ever believed earth to be the most important place in the cosmos. That was heaven, God's dwelling place. Between base, corrupt, contaminated earth down here and heaven up there stretched a linear hierarchy of celestial spheres and pristine stars, unchanging and perfect.

Dedicated to the Pope, Copernicus's *On the Revolutions of the Celestial Spheres* isn't put on the list of banned books, until well into the seventeenth century, which is when Protestant sects seize on heliocentrism as a liberating theology.

If the earth has a diurnal rotation and an annual orbit, they argue, that means that Christ's illuminating light shines equally on us all. The light of divine inspiration touches the laity directly. What price linear hierarchies now? Who needs bishops, kings or priests? We are all saints. Pamphlets with titles such as *More Light Shining In Buckinghamshire* argue that heaven might not be a place up in the sky, beyond the Celestial Spheres, but instead heaven could be built in the here and now, and could actually be a place on earth. The Counter-Reformation suppresses this idea, but it resurfaces in the work of Belinda Carlisle.

When it comes to evolution, Freud is a notoriously unreliable witness. He goes to his grave arguing against natural selection because Darwinism contradicts Freud's own theory of evolution: phylogenetic recapitulation. Freud's theory of evolution deserves to be much better known because it is gloriously nuts. Each individual recapitulates in one lifetime the entire development of the human race. As babies we crawl on all fours like when we were monkeys come down from the trees. Then we learn to walk upright and develop rudimentary language just like our ancestors in the Pleistocene epoch.

Neurotics, says Freud, are people who have got stuck at a previous historical epoch – such as the last Ice Age, when a friendly world turned suddenly forbidding, and dwindling resources restricted sexual activity. Or the Bronze Age, when a tyrannical clan chieftain was slain by his sons who felt great elation and liberation but also terrible guilt at their bloody deed. 'Triumph over his death', writes Freud in *Totem & Taboo: Some Points of Agreement Between the Mental Life of Savages and Neurotics* (1913), 'is then followed by mourning over the fact that they still revered him as a model.'

Freud offers his theory of evolution as a *diagnostic tool*. If you're a psychiatrist and a patient presents with anxiety, paranoia and sexual impotence, you know they are reliving the Ice Age. If someone is suffering wild mood swings accompanied by terrible guilt, they are reliving the Bronze Age. If someone is singing 'Land of Hope & Glory' while stabbing their own genitalia, they are reliving the Brexit vote.

For what it's worth, few of the great scientists considered their own work to be delivering a hammer blow to humanity's naive self-love. Darwin concluded the *Origin* with the ringing declaration: 'there is a grandeur in this view of life'.

Despite Freud's shaky grasp on the history of ideas, his thesis has become a cliché of popular thought and set the tone for the nihilistic style in contemporary science writing. If the hallmark of great scientific advances is to make everyone feel worse about themselves, the science writer reasons, then if I tell people they are shit I'll be hailed as a great scientist!'

Maybe that's what they are thinking, maybe it's not. But what is clear is that science writers now compete over who can say the most horrible thing about the rest of us.

'The human race', said Stephen Hawking, 'is just a chemical scum on a moderate-sized planet, orbiting around a very average star in the outer suburb of one among a hundred billion galaxies.' Not to be outdone, Yuval Noah Harari would have you know that: '[t]he free individual is just a fictional tale concocted by a set of biochemical algorithms.'

This tough talk delivers cold hard reality for those who can handle the truth – which is ironic really, because Hawking's picture is pure *Alice In Wonderland* because he hasn't thought it through.

For us to look like scum to an alien observer then this alien has to know about scum. The alien has to have a concept of scum. It is impossible to have a concept of scum without knowing about living organisms. The aliens must be able to distinguish scum from organism. Scum is dead; organisms are alive. The spirogyra wiggles the scum. The scum does not wiggle the spirogyra. Spirogyra, water fleas and humans may swim through the layer of dirt or froth on a liquid but are not part of the scum. All of this must be known to the alien observer before that alien observer can know what scum is. If, however, these aliens persist in classifying us as scum on the grounds that they are so immeasurably superior to human beings as to make any distinction between life and death negligible, then I suggest that it is they and not us who are scum.

An argument about human insignificance that relies on anthropomorphism is surely self-defeating. The anthropomorphism is of course the term 'outer suburb.' But if we are the outer suburb, then where is the bustling downtown hub? The centre of the galaxy? But that is as dead. There is no life there. Hawking surely knows the Galactic Centre to be dead, and so it is hard not to question whether he really believes what he is saying.

This is a characteristic feature, by the way, of the whole sadistic, macho genre of modern science writing. The science writer only wants to be *a little bit* nihilistic, but never goes the whole Raskolnikov, which leaves him in a mealy-mouthed limbo land, insinuating that all is cold and hollow but not wanting to scare off the middle-England fan base, many of whom are members of the RAC.

Wearing a double-breasted blazer with brass buttons, Richard Dawkins announces that we are vehicles for genes, but then ducks the logical entailments. If genes are in control, then all emotion is not just contingent but illusory. One of the many nettles Dawkins never grasps is that this abolishes love. By his own logic love is no longer a contingent fact of natural selection, but illusory. And yet he would never do a book called the *Love Delusion*. That might upset Mrs Dawkins. * And it would come over as a little bit *too* nihilistic, and that would never do when one has tickets to sell to retired Rotarians at Hay-on-Wye. Now, me personally, I don't believe in a loveless world, but that is the logical entailment of his hypothesis that clever genes get us to do things they need us to do, if you think it through.

---

* Thanks to Simon Munnery for this analogy.

Trash-talking the human race rides an executive car that whooshes through an automatically raised barrier with no inspection of documents necessary. You do not need to do anything so otiose as actually to prove that we are selfish sociopaths who live in a world of illusion. That's axiomatic. That's where you start from. Evidence that people are deluded in all but their reptile brains does not need to be proved yet again before theorising that smiling is snarling or any of the thousand other slanders and libels in any of the thousand other brain books currently sagging bookshop shelves. Why waste time proving what science has already established? Science has found people bad. Do we have to prove sea water to be salty each and every time we submit a paper on the Atlantic Ocean? No. It's understood. Seawater is salty. People are bad. Science has proved it. Let's not reinvent the wheel. Let us proceed from axioms. Let the burden of proof fall only on those with a good word to say about humanity. Let those Pollyannas who want to drag us back to a pre-empirical mysticism prove their case. Until then, contempt is the one and only scientifically correct position from which enquiries into the deluded human race should proceed.

This nihilistic turn in science writing has gravely damaged intellectual life by lowering the threshold of proof for any condemnatory claims about human nature, whilst at the same time raising the threshold of proof for anyone with a good word to say about us. Now all this damages more than just intellectual life, of course. To tell us we're nothing, to say that everything we see is an illusion, disempowers and dismays us. It also affects how we look at each other. If we see people struggling to control their lives, they suddenly seem to be behaving more or less irrationally.

When it comes to scientific investigation, the nihilistic take pre-emptively shuts off interesting avenues of inquiry. A generation ago, for example, the Time and Motion Men swept through evolutionary biology subjecting all human experience to cost-benefit analysis.

The Time and Motion Men

In the early twentieth century, Time and Motion Men were sent into factories, warehouses and offices to rationalise every moment of the workers' day to make it as cost-beneficial as possible. They measured the distance from lathe to toilet, counted how many seconds were unnecessarily wasted in reaching for hammer or in saying hello to colleagues. Everything superfluous had to go in the belief that this would produce a more efficient work place.

In the late twentieth century, the Time and Motion Men decreed that every evolutionary adaptation had to be shown to have passed a cost-benefit analysis. They tried to account for the selective advantage of everything under the sun. The mobula ray, a flying fish, seems to leap into the air for the sheer joy, the exuberance of taking it to the air and then landing with a big splash back down in the water. Sheer joy? No, every adaptation has to earn its keep. There has to be a practical reason. And so the Time and Motion Men argued that reason nature selects for mobula rays that leap into the air is because this activity makes them more athletic, and better able to avoid predators. But naturalists and logicians pointed out that they still had to account for why these mobula rays started leaping in the first place. Flying fish had to be flying already for flying to be selected for.

How could nature select for mobula rays that leap into the air if mobula rays weren't already leaping into the air? The problem of joy was a terrible headache for the Time and Motion Men. It made their lives a misery. No matter how far back they went an animal having fun had always got there first. An elephant was always to be found joyfully rolling in mud long before nature selects for the parasite repellent qualities of a mudpack. The sad truth was that the elephant was doing it because it felt brilliant. And thus the Time and Motion Men fell down the Well of Infinite Regress.

I want to emphasise that infinite regress is a problem only for the cost-benefit story of selection. It is *not* a problem for Darwinian natural selection. For Darwin, animals have a physical, emotional and psychological constitution. Mobula rays give full play to their emotional constitution when they break water and flap their fins in the sky. Elephants follow their pleasure when they roll in a mud bath. Nature selects for or against this constitution, for and against the entire the life cycle that goes with it.

If joy dismayed the Time and Motion Men, language completely tormented them. It was a standing provocation. Why would people just give away information for no immediate return? It didn't add up. Where was the competitive advantage in that? Where was the cost-benefit? Extraordinarily convoluted reasoning tried to square the emergence of language with best economic practice. They embarked on a prolonged filibuster about game theory, but this seems finally to be petering out. In the last few years less narrowly dogmatic ideas about the origin of language are getting a hearing, ideas that take us back, in fact, to a long-neglected insight of Charles Darwin's, which we will look at in the next chapter.

12. LIKE YESTERDAY

There is a hard and fast convention about how scientific epiphanies are supposed to happen. The scientist should be all alone, deep in thought, perhaps in an empty late-night lab, and despairing of ever solving an intractable problem. But when Charles Darwin had his epiphany about the origin of language he was on all fours with an apron covering his face.

On 27 December 1839, after the birth of his first child, Darwin began taking detailed notes on the development of the boy William, nicknamed Doddy. In these notes, which he called 'A Biographical Sketch of an Infant', Darwin records the exact date at which Doddy first shows signs of the human behaviours, such as: *'Anger. Fear. Affection. Association of Ideas, Reason, & c. Moral Sense. Shyness. Means of Communication'.*

The purpose of these notes is to compile empirical data about what is instinctual and what is acquired in newborns. They are also an accidental portrait of a besotted father, woozy with oxytocin, and full of love for his first-born.

He lists all the things that make Doddy laugh, such as when he covers his face with an apron and then whisks it off. Over and over again, Darwin whisks the pinny from his face, eliciting squeals of joy. He throws shadows on the wall, pleads for a kiss, growls like a monster, pulls faces when they are both sitting in front of the mirror, and notices that precise date when Doddy first looks round from the reflection to him.

In making these notes, Darwin is on the lookout for universal milestones in infant development. What stands out, however, isn't the universal but the particular. What we read are notes on a very Victorian, very English infant of the *haute bourgeoisie* – as the following entries show.

When five months old ... as soon as his hat and cloak were put on, Doddy was very cross if he was not immediately taken out of doors.

At 6 and half months, he displayed violent passion when dressed in clothes he had already worn that season.

At 11 months Doddy became exceedingly distressed upon hearing nursemaid express support for Irish Home Rule.

At 14 months, after suckling at the bosom, burst into tears of remorse, and spoke his first complete sentence: 'I have polluted your purity, Madam, with my depraved lusts and vile appetites!'

At 18 months, while nursemaid reads to him, Doddy becomes agitated by rise of literacy among the lower classes. Nursemaid manages to calm him down by saying that she memorised the story through an oral folk tradition, and just has the book there as a prop which she is only *pretending* to read, as she is completely illiterate. Doddy was much assuaged by this very clever answer. A bit too clever, if you ask me. Will fire her tomorrow. Let her find work with her Fenian socialist friends!

Some observations, however, do have the universality for which Darwin was probing. In the section entitled *Moral Sense*, for example, he finds an answer to a question which has long fascinated students of psychology: at what age do we first deliberately and consciously deceive? With Doddy this comes when he is 2 years and 7½ months old, which is when Darwin:

> … met him coming out of the dining room with his eyes unnaturally bright, and an odd … or affected manner, so that I went into the room … and found that he had been taking pounded sugar, which he had been told not to do … A fortnight afterwards, I met him coming out of the same room, and he was eyeing his pinafore which he had carefully rolled up; and again his manner was so odd that I determined to see what was within his pinafore, notwithstanding that he said there was nothing and repeatedly commanded me to 'go away', and I found it stained with pickle-juice; so that here was carefully planned deceit. I informed him that from now on he was dead to me.

But what snags Darwin's attention more than anything else is the musicality of Doddy's speech. 'The use of these intonations seems to have arisen instinctively', he notes. The more he listens to Doddy's cadences, the more he wonders whether language evolved from a combination of singing and mimicry. The first germs of Darwin's origin of language theory begin to form. His theory of language can be stated in this way: first came the music and then came the lyrics.

With typical Darwinian alacrity he mulls over this theory for the next thirty years before rushing into print. This is the man, after all, who spent eight years writing a book about barnacles. (You don't want to be too hasty about barnacles.) He was, if nothing else, a man tuned to geological timescales, who slowed his perception of the world until he could see the worms churning the soil and lifting the fields by inches.

The 1870s are his decade of publishing ideas about specifically human evolution, and this is when he develops his origin of language theory. In *Descent of Man* (1871), he argues that 'some early progenitor of man probably first used his voice in producing true musical cadences, that is, in singing.' Language came from mimicking 'the voices of other animals', and from modifying our 'own instinctive cries, aided by signs and gestures'.

The following year, in *The Expression of the Emotions In Man and Animals* (1872), the origin of speech in song is narrowed down to love songs and serenades. After noting how gibbons sing 'an exact octave of musical sounds, ascending and descending the scale by half-tones', Darwin writes that our ancestral primates: 'probably uttered musical tones before they had acquired the power of articulate speech', and used these musical tones for wooing, and so they 'became associated with the strongest emotions … ardent love, rivalry and triumph.' A vestige of this operatic origin of language, he argues, is the way the voice gets more musical the more emotional we become.

In 1877 Darwin submitted 'A Biographical Sketch of an Infant' to the journal *Mind: A Quarterly Review of Psychology and Philosophy*. After first describing how Doddy's 'whine of dissent had a different resonance and timbre' from his 'strongly emphatic ... humph of assent', Darwin concludes that the instinctive sing-songs of infant speech allows us to infer that:

> before man used articulate language, he uttered notes in a true musical scale as does the anthropoid ape Hylobates [gibbon].

The idea that language emerged from singing seems frivolous set against the dog-eat-dog tone prevalent in evolutionary theory.

In the classic work *Language: Its Nature Development and Origin* (1922), Otto Jespersen says that language theorists tend to imagine 'our primitive ancestors ... as sedate citizens with a strong interest in the purely business and matter-of-fact aspects of life.' Jespersen argues that 'the genesis of language is found ... in the poetic side of life; the source of speech is not gloomy seriousness, but merry play and youthful hilarity.'

Though the insistence on dourness in theories of language is nothing new, it became more pronounced after the Time and Motion men came calling on evolutionary theory.

The Time and Motion men may have disappeared down the Bottomless Well of Infinite Regress, but their influence is everywhere, not least in the fact that Darwin's theory that language evolved from singing is not widely known. But that may be about to change. Lately, two lines of enquiry have converged upon Darwin's origin of language theory, one of which we can call Colossal Hypoglossals and the other The Paleoloithic Crowd Sourcing of Language.

Colossal Hypoglossals

In human skulls the anterior condylar canal is a groove scored by the hypoglossal nerve that articulates the tongue. The stronger and more dexterous the tongue the fatter the nerve. In ape and *Australopithecus* the anterior condylar canal is tiny. In endocasts of *Homo heidelbergensis* skulls, paleontologists have discovered the condylar canal to be exactly the same size as our own. These colossal hypoglossals show that our prelingual ancestors could do all the noises we do. Any sound we can make they could make. Hundreds of thousands of years before spoken language, they could perform alliterative tongue-twisters and assonant rhymes. They could do funny voices to stop the children crying, or mimic forests sounds to distract the dying from their pains. They could impersonate all the subtle gradients of sound that soup makes as it comes to the boil. And they could sing.

It would appear that cave men and women were not going 'Ug Ug Ug' after all. When producers of TV science documentaries reconstruct our hominin ancestors, their stunted vocabulary (hominins, not TV producers) always goes with stunted vocal range, the unlikely inference being that the vocal chords becomes more supple the more words you know, as if the best singers were the most articulate conversationalists. Is it time to start imagining our forebears as less butch? Perhaps we could begin this process by renaming early humans.

At present we name our ancestors after remote parts of Germany, such as the Neander Valley or Heidelberg – the Germans having got their bones, like their swimming towels, down early. Or else we name them after the most popular types of tools or hardware, be it hand-axe or beaker. It is a sobering thought that when our bones are dug up in a future epoch, we may also be classified according to our most popular hardware. The B&Q people, the Wickes Man, the Robert Dyas of the Younger Dryas.

Early humans were more likely to have identified themselves by musical genre than by hand-axes. If only acoustic caves existed that could somehow preserve ancient sounds, then we might more accurately name early humans by musical genre. *Homo emo, Homo mbaqanga, Homo milonga, Homo gamelan, Homo Nederpop, Homo Happy Hardcore,* The Old Wave of Old Wave people. The *A Capella* people.

Naming hominin groups after their distinctive musical genres might also throw light on intractable mysteries of prehistory. It could have revealed to us, for instance, that early humans left Africa due to musical differences.

The Crowd Sourcing of Language

We were living in complex communities for at least half a million years before we started talking to each other. A growing school of thought argues that a rich non-verbal human culture came first and language a distant second. There were complex interactions long before there were words to help them along. Or to unnecessarily complicate them, as the case may be. In this chapter I want to explore the idea that language was the first crowd source drawing on culture – the first cloud store.

In *A Mind So Rare,* Merlin Donald reverses the usual scenario where the early humans first evolve language and the use it to create a culture: 'Just as the physical environment drove the evolution of perceptual capacities, so cultural energy drove the evolution of sophisticated communication capacities.'

I like the general tenor of this, but I'd just like to add a caveat that the physical environment can also drive cognitive capacities. In my last book *The Entirely Accurate Encyclopaedia of Evolution,* I looked at how more complex coral produces more intelligent fish. Of two populations of stripy blue cleaner fish on the Great Barrier Reef, the population living in coral, with more nooks, crannies, and biodiversity, are more intelligent than those inhabiting a duller stretch of coral.

In their 2014 *Ethology* paper, 'Variation in Cleaner Wrasse Cooperation and Cognition: Influence of the Developmental Environment?', Sharon Wismer *et al.,* argue that the more complex physical environment elicits the phenotype of smarter fish among a population of stripy blue cleaner fish. They describe cleaner fish so intelligent that they recruit moray eels to their hunting parties, with nods of the head. In calm water moray eels use the lid of the sea as a mirror. By looking up at the glassy ceiling they can see who or what is hiding behind the coral. They also use this aquatic periscope to see round corners.

Sharon Wismer and colleagues found that cleaner wrasse from larger shoals are more intelligent than those from smaller populations. So it appears that both physical and social complexity make intelligent fish, which brings us back to Merlin Donald's point about the social basis of the inner life, how language couldn't have emerged except as a product of living in complex groups, since it exceeds what any one person was ever equipped to do on their own. Long before early *Homo sapiens* had language, he says, they were able to 'bond, coordinate group activity, transfer and refine skills, and establish a network of custom and convention.'

This chimes with ideas of the social basis of thinking and speech advanced by the brilliant Russian psychologist Lev Vygotsky (1896–1934), who died of tuberculosis when he was just 37 years old, but whose work is currently enjoying a renaissance.

Fifty years after Darwin finally published his observations on Doddy's early learning, Vygotsky and his team of researchers were combining detailed observation of early learning with imaginative experiments designed to explore the different ways in which children solve problems alone or in groups, the ways they talk to themselves, their use of whole phrases and chunks of borrowed speech, and how these are deployed when performing tasks within and then beyond their competency, with and without help from adults.

Until Vygotsky the following tale was told about how our thought and speech develop. (And since Vygotsky, too, alas, so much did his early death damage his influence.)

A child starts off with her own internal monologue which she then takes to nursery school where she learns to turn monologue into dialogue, and learns to merge and to attenuate the imperious demands of self with the recognition that there are others, and that those others possess, in the words of George Eliot, an 'equivalent centre of self.' Vygotsky came to believe that what actually happens is precisely the opposite. Through being with others – family, nursery, reception – infants learn speech and thought, which they then use to pollinate a rich inner life.

What we can do in groups excels what we do alone. Development traces an intricate dance between the inside and the outside world. When Vygotsky says that a five-year-old's experiences shape her brain more than her brain shapes her experiences, he is absolutely not arguing that the child is a blank slate. It is the very richness of our emotional constitution that demands so much external nourishing. If there were less to us, we wouldn't need so much outside assistance and support. As the philosopher Mary Midgley puts it:

> ... society is not an *alternative* to genetic programming. It requires it. To become a member of any kind of society, an infant must be programmed to respond to it. Others give him his cues. But he has to be able to pick them up and complete the dialogue.[*]

The greater the endowment the more there is to learn.

This learning takes place in what Vygotsky called the 'zone of proximal development'. This zone describes those capacities just out of reach. The zone is the shimmery no man's land between what the child can almost-but-not-quite do on her own, and what she can do with the teacher leading her. Good education focuses on this zone. Vygotsky developed these ideas in the 1930s. His ideas should not be confused with the contemporaneous teaching methods being developed at Marcia Blaine School for Girls in Edinburgh, where Miss Jean Brodie told her pupils: 'The word "education" comes from the root *e*, out, and *duco*, I lead. It means a leading out. To me, education is a leading

---

[*] Mary Midgley, *Beast and Man: The Roots of Human Nature*, 2002.

out of what is already there in the pupil's soul.' Education is that, of course, but it's also the making of the pupil's soul.

If not like Jean Brodie's, Vygotsky's ideas are close in sprit to those of another contemporary pedagogue, American philosopher John Dewey, who in 1929 wrote:

> Failure to recognize that [our] world of inner experience is dependent on upon an extension of language which is a social product and operation led to the subjectivistic, solipsistic and egotistic strain in modern thought.

The social basis of consciousness allowed others to be present for us even when not actually physically there, because far away or dead. 'The words of the dead', wrote Auden, 'are modified in the guts of the living.' This modification was the germ of the inner voice. Long before writing, repeating an elder's rhythmic mime when we were alone helped us recall where the fish nets were sunk, or helped us remember the correct sequence of actions needed to gut a marmot, or start a fire. This repetitive mime could have been done by actions and a work song with no words, only noises. This would have been a type of prelingual transmission of information and encouragement by way of movement and song. Slowly and gradually, early humans crowd-sourced language from the cognitive communal cloud store. Such theories of language, and the discovery that colossal hypoglossals scored deep grooves in *Homo heidelbergensis* skulls, lend support to the epiphany that first struck Darwin when playing with his first-born son, his theory that language evolved from 500,000 years of singing.

Our genus *Homo* is two million years old, but language emerged only forty thousand years ago. This is like yesterday. It's also like 'Yesterday'. Famously that song's tune popped out perfect and fully formed (in a dream Paul McCartney had at Jane Asher's house), but the words were a long time coming. According to John Lennon, 'the song was around for months and months before we finally completed it.' 'Yesterday' was originally 'Scrambled Eggs' and its opening couplet went like this:

Scrambled eggs,
Oh my baby how I love your legs.

But the words didn't take because they didn't fit the emotional timbre of the music. Just as complex, lyrical expressions gradually emerged to turn 'Scrambled Eggs' into 'Yesterday', so, says Darwin, 'the most complex and fine shades of expression must all have had a gradual and natural origin.'

If Darwin is right, if we communicated by music tones before the invention of speech, then perhaps instead of names everyone had a leitmotif, like in *Peter and the Wolf*. I mean we must have had some way of naming each other before the invention of language some forty thousand years ago.

A dissenting voice to all this is my own theory of the origin of speech. After a thirty-year long stand up comedy career, I like to think I've become something of an authority on long uncomfortable silences. After the first quarter of a million years without talking, I suspect early humans couldn't stand the silence anymore. The long silence started to get really uncomfortable. People couldn't bear it. Something had to be done. Of course we didn't burst into speech straight off. Instead I think we broke the eggy silence with some tentative beginnings. It was a gradual process. Language began, I believe, with a few thousand years of strained humming and self-conscious throat clearing. Towards the end of the Late Pleistocene, it seems probable that 'ahem-ahem' may have given way to someone rubbing their hands together after a single clap, and that was when, in an overly loud voice, the first ever human word was spoken, and it was: 'Annn-yy-way…'

13. THE MYTH OF THE STONE AGE BRAIN

The Myth of the Stone Age Brain has become a staple of modern thought, the central dogma of self-help books and landmark TV series, a touchstone of psychology, neuroscience, evolution and anthropology. It has been used to explain everything from sub-prime mortgages to childhood obesity, and from xenophobia to the global epidemic of mental illness. It's the idea that our Stone Age brains are bewildered by the modern world. We are refugees from the Pleistocene watching a fast-forward world hurtle past us, and 'suffering from centuries of taming', as I believe Adam Ant put it in 'Kings of the Wild Frontier'. One minute we're knee-deep in a Pleistocene lake squabbling over fish-kill with Great White Pelicans, and the next thing we know we are up to our necks in online tax returns. No wonder we can't cope.

Recent findings that the Late Pleistocene was a time of radical instability, due to a concatenation of extreme climatic events, have not been enough to get people to give up their idea of the Pleistocene Pastoral Idyll.* Instead science writers carry

---

* Mark A. Maslin *et al.*, 'East African climate pulses and early human evolution', *Quaternary Science Reviews*, 2014.

on dreaming of those the halcyon carefree days on the savannah. 'Our brains are adapted to that long-vanished way of life', sighs Steven Pinker in *How The Mind Works*, 'not to brand-new agricultural and industrial civilizations'. Modern life, it seems, is a terrible wrench:

> [Our brains] are not wired to cope with anonymous crowds, schooling, written language, governments, police, courts, armies, modern medicine, formal social institutions, high technology, and other newcomers to human experience.

Is 'cope' the right verb to describe our relationship with the written word? Is reading all too much? In what sense, for goodness' sake, do we cope with menus and sign-posts, sheet-music and score boards? And does 'cope' capture our relationship with modern medicine, schooling and anonymous crowds? I mean, sure we have our problems, but these tend not to be Calpol, double P. E. and Glastonbury. Are we really so very brittle? Has cultural evolution been so frightfully onerous?

Louise Barrett, author of *Beyond The Brain*, questions the odd passivity in the version of humans that Steven Pinker gives us here. 'But who invented those things?' she asks. 'We did. How can you invent something and then not understand it? How can you not cope with them if you made them in the first place?'

While it is possible to invent things with which you find you can't cope – an economy based on burning fossil fuels springs to mind – human culture is not a Frankenstein's monster with which humanity cannot cope. It is what has enabled us to cope. It is what we made of things. This seems to me a pretty uncontroversial conclusion, with which most people would agree. And yet incredibly, Steven Pinker's mawkish, sick-note view of the human mind and human culture has swept all before it. So I think it is well worth going through each item in Pinker's sick-note, and to examine his litany of complaints one by one.

Anonymous crowds

To say that our brains can't cope with anonymous crowds is to say that agoraphobia is the norm. But one of the things that makes agoraphobia so tormenting for its sufferers is precisely that they are unable to share in the ordinary pleasure that every-one else takes in festivals, dance halls, theatres, markets, fun runs, pilgrimages and weddings. This means they miss out on important rites of passage involving family and close friends.

Now, if someone is going to come along and claim that our brains are wired for a specific mental illness – such as agorapho-bia – then you might expect them to have strong corroborating evidence. I mean, it's a pretty large claim, after all. But we get no evidence whatsoever. We are just supposed to accept this sour view on trust. (Which presupposes that we are either very sour or trusting. One or the other as the two don't usually go together.)

When it comes to understanding human behaviour, this mawkishness puts us on completely the wrong foot. Only a perfect stranger to humanity could fail to notice the delight we take in large gatherings.

Humans seek out other humans. Young people especially have always gone out of their way to get to know exactly the sort of people they didn't grow up with. Sometimes this wanderlust was formalised, as in the ancient Aboriginal Australian rite of passage called 'walkabout', where young males follow 'songlines' and go on off on their own for six months, passing through other territories, and encountering other tribes along the way. All cultures recognise the fledging instinct in the young. There is no consensus among different cultures over what is to be done about the urge to fledge. In some cultures it is indulged, in others strictly forbidden, albeit at the risk of the young people running away from home to join another community. But all cultures agree that it is a force to be reckoned with, in one way or another.

So much for individual wanderlust, what about groups of people? If a person was ever exiled from one settlement and went to live in another, if raiders ever carried off captives for slaves, if a foraging party ever wandered round lost until it entered the territory of another tribe – if any of these things ever happened then adapting to anonymous crowds is nothing new, and precedes agricultural civilisation by a very long chalk. For his part Steven Pinker happens to be an advocate of the 'Paleolithic-human-warfare-hypothesis', the creed that fifteen per cent of prehistoric populations were slaughtered in wars. For these to have been actual wars, and not just dust-ups, not just a couple of tooled-up fellers intent on homicide, means an anonymous crowd appearing over the ridge at dawn.

So much for groups of people, what about those anonymous crowds called migrating populations? Humans as a species prefer mountain pass to mountain top, river bank to exposed plain, coast to sea. As early human populations migrated across vast territories, they discovered pathways, footprints, and each other. For our genetic traits to be shared so widely among us, then bands of nomadic foragers must routinely have bumped up against other bands of nomadic foragers. For proof of which, we need only consider the wide distribution of Upper Paleolithic Venus figurines.

Venus In Furs

About 35,000 years ago, while stuck inside waiting for the last Ice Age to end, people began carving voluptuous Venus figurines, all hips, belly, buttocks and breasts. Like most primitive art, Venus figurines possessed magical powers. Venus worked her magic from Gagarino in Ukraine to Parabita in the heel of Italy, from Hohe-Fels in Germany to Laussel in the Dordogne. We don't know whether Venus trinkets were given as gifts to protect travellers, traded for provisions, or sold as tour merchandise at Venus gigs. They may have been ceremonial gifts exchanged in return for safe passage, or in return for being able to share the host tribe's winter larder. These gifts 'allowed people to navigate their way through critical negotiations'* – critical because if they didn't come off, wandering tribes would not be allowed to over-winter in the host tribe's warm cave complex. If negotiations fail, you starve to death in the ice. But thanks to the offering of a magical Venus, nomadic foragers were allowed to hunker down in the last Glacial among strangers.

One way or another, those who carried or copied, bought or sold, traded or stole a magical Venus must have come across crowds of people whose names they didn't know. If the experi-

---

* Christopher Stringer & Clive Gamble, *In Search of the Neanderthals*, 1993.

ence didn't make them sterile with shock, if they were still able to have babies, then meeting anonymous crowds must be part of our genetic inheritance. Through our bodily comingling we shared parasites, germs, sexual fluids and acquired disease resistance. Gregarious existence selected out those unable to cope mentally with anonymous crowds. Those who suffered a panic attack and fled gibbering into the ice and snow would tend to be less successful in producing descendants who shared copies of their catatonic genes.

Even if some populations never saw a crowd of people until the dawn of agricultural civilisation, that still gave everybody a fair few millennia to get used to them. It's not as if anonymous crowds suddenly snuck up on us. If the Neolithic marked the rise of small-scale agriculture, it was also a period of getting used to large crowds. For proof of which, we have that world-famous standing testament to the phenomenon of anonymous crowds: Stonehenge.

The sheer labour of hauling the bluestone from the Preseli Hills to Salisbury Plain would have taken a small army. What else would this vast work-party be to the settlements they rolled through but an anonymous crowd?

Stonehenge was built by and for people who came from far and wide to gather in one place. We don't know why – maybe for the healing properties of the bluestone's dolerite and rhyolite. Maybe they got a buzz from seeing lots of people whose names they didn't know, or a contact high from dancing and singing in a big crowd of strangers. But surely not even the most militant trickle-down free-marketeer believes Stonehenge was built as a private residence? A stately pile for just one wealthy couple? On rainy nights, snug beneath its roof of hide, these exclusive freeholders would look out the Heel-Stone Window over the wind-hammered hovels and tumbledown benders on Salisbury Plain, and say: 'I refuse to feel guilty. I worked bloody hard for this. If those people on the plain had spent a bit less time drinking woad and dancing round the fire pit, they could be living in a place like this themselves!'

If our brains are not wired to cope with anonymous crowds, how comes it that they do? Do we endure anonymous crowds by dint of unconscious repressions which exact from us a greater psychological toll than we realise? ('Doctor, I feel so tired all the time.') There is a lot of uptake for this sort of explanation. Everyone likes to pose as the outsider. But it still leaves unexplained the unmissable fact of human gregarity, our delight in anonymous crowds.

In *On Going A Journey* (1822) William Hazlitt describes the first time he ever set foot in France. He travels alone, speaks no French, and steps off the gang plank slap bang into an anonymous crowd:

Calais was peopled with novelty and delight. The confused busy murmur of the place was like oil and wine poured into my ears, nor did the mariners' hymn, which was sung from the top of an old crazy vessel in the harbour as the sun went down, send an alien sound into my soul. I only breathed the air of general humanity.

How he bears up under this torment I'll never know. It's a testament to human fortitude if nothing else. The phrase 'confused, busy murmur of the place' sends shivers down the spine, but Hazlitt, strange to say, seems actually to relish the experience. Whether by dint of iron constitution or genetic mutation, he also – somehow – copes with alien sounds being sent into his soul. Freak.

Schooling

Who taught the Lascaux cave painters to draw? Where are their crossings out? Why aren't there any of those buffaloes where halfway through, you realise you've overdone the shoulder hump and you have to pretend that you were trying to draw a camel all along? Unless of course the Lascaux artists actually started off trying to draw a camel but halfway through, realising they'd got the hump too far forward, pretended they had been trying to draw buffaloes all along.

Governments, Police, Courts

The idea that the rule of law – governments, police and courts – is an imposition alien to our nature has been around for a very long time. It is an idea which for Bertrand Russell shouldn't have outlived Darwinism: 'The theory that government was created by a contract is, of course, pre-evolutionary. Government, like measles and whooping cough must have grown up gradually.'* For Darwin, as for Russell, our social instincts lead us to create all kinds of elaborate rules and laws. Our natural inclination to live in society entails complex protocols.

For as long as there have been humans there have been dos and don'ts, no-nos and protocols, crimes and punishments. And

---

*   Bertrand Russell, *A History of Western Philosophy,* 1945.

for the same amount of time there have been those who find these irksome and dream of living like kings of the wild frontier, as wild and free as animals on the savannah. The problem is that animals in the savannah such as African hunting dogs and hyenas are also bound by all kinds of rules accompanied by sanctions and punishments. Mammals who want to live a lawless life must live alone, as Siberian tigers and golden hamsters do.

As for our brains' inability to cope with police, *The Guardian* has argued that it is the other way round: the police are struggling to cope with our brains. The College of Policing in the UK, the paper reports, estimates that between 20 and 40 per cent of police time is spent dealing with the mentally ill, while *The Guardian*'s own research discovered a 33 per cent rise in cases linked to mental health, over the three years from 2011 to 2014.[*]

There is much debate over how much of this increase is due to cuts in mental health provision and how much is due to a nosedive in people's mental health, but there is no debate – quite rightly – about whether this is due to Stone Age brains struggling with a modern world. There is an excellent reason why this is not seriously debated. It is because the 33 per cent rise in crimes isn't since the Pleistocene but since 2011.

* * *

If our brains aren't wired for courts, then how do juries and barristers, judges and ushers survive it every day? What happens when the brains of the legal profession at last give way to that innate

---

[*] *The Guardian*, 27 January, 2016.

law-court phobia of which is our collective genetic inheritance?

DEFENDANT: Well it is certainly a great comfort knowing that you will be by my side during the trial.

BARRISTER: Actually, I shall speaking by video link from a beach hut in Whitstable. I just can't cope with the Old Bailey. I don't know if it's the smell of polished oak, the sound of swishing robes, or just knowing that there's criminals like you literally inches away. No offence.

* * *

Human beings find injustice intolerable. Formal processes of third-party arbitration and public restitution – courts of law – are found in all cultures at all times. Sometimes these formal processes happen in a special place separate from the rest of life, such as the Lögberg (or Law Rock) of the ancient Icelandic parliamentary council the Thingvellir. These formal processes often involve ceremony, or some kind of ritual purification, the better to attune yourself to the demands of the inherited community of laws, or the better to hear the whisper of an oracular goddess. The Greeks set aside a *temenos*, a sacred place dedicated both to the official business of the city-state and, when humans could not agree, to prayers for divine intervention.

A special language may be spoken or special customs enacted or a special title given so as to ensure that the actors in the legal drama are aware that they are no longer solely an individual but a representative of a quorum, and duly undertake their solemn duty, free of distraction. In Navajo law, for example, the *naat'aa-*

nii or peacemaker hears conflicting cases and then administers restorative justice.

None of these elaborate rituals are entirely alien impositions. All grew out of the deeply human need for the recognition of wrongs, for restitution, for justice. I'm not saying they reflect the needs and wishes of all people at all times. Property rights frequently prevail over human rights, for example, and as Bismarck said: 'Those who love sausage or the law should not enquire too closely into how either are made.'

But if not government, police or courts, what is this thing that was supposed to have suited us so much better? The arbitrary justice of a warrior caste? It is hard to think what else served that function of the rule of law in a world without government, police or courts, than a warrior caste. And this brings us to the next item on the list.

Armies

Now here at last Pinker does have truth on his side. People really do find it difficult to cope both with being visited by an army and being in an army.

As soon as you have a warrior caste, you have a group whose ethos and interests are in important respects different from those of the host society. Nothing so unusual about that of course. It's also true of craft guilds or religious sects. The difference, though, is that the warrior caste are so very much better at violence.

A good illustration of this is provided by philosopher Mary Midgley in her essay 'On Trying Out One's New Sword'. She explores the Samurai warrior's sacred duty of *tsujigiri*. In *tsujigiri* tradition, the Samurai must test his new sword and the test material should be a random traveller. There was really no other way to know how well-tempered the steel was, or how grippy the handle than to cleave a stranger in two. *Tsujigiri* translates as 'crossroads cut' – a cut made at the crossroads. Samurai would sit at wayside inns by major intersections waiting for a chance wayfarer to become his sword test dummy. Eventually local innkeepers demanded an end to *tsujigiri* because it led to terrible ratings on TripAdvisor: 'Staff friendly. Miso soup tasty. Husband's arm lopped off during check in.'

Modern medicine

Modern medicine may indeed be a newcomer to human experience but, as Stephen Jay Gould points out, it has given us something else entirely new as well:

> [T]hanks to modern medicine, people of adequate means in the developed world will probably enjoy a privilege never vouchsafed to any human group. Our children will grow up; we will not lose half or more of our offspring in infancy or childhood.

How are we coping with that one? How are we bearing up under the strain of not burying our own children? This shows how the Romantic idea of a Stone Age brain bewildered by modern world is possible only so long as you exclude what bewildered our Paleolithic ancestors. Such as not knowing how to keep the children alive. Parents then as now were dismayed by how hazardous and uncertain life is. Bewilderment is nothing new.

How much were our brains 'wired to cope' with prehistoric medicine? As a headache cure, paracetamol agrees with us in a way that trepanning (the drill-hole cranial surgery of yesteryear) never really did. Paracetamol bonds with the TRPA1 protein found on the surface of neurons. Our liver might not cope, but our brain does.

The Willing Mind

Pinkerism is the mawkish belief that if something is difficult it's difficult to the extent that it's a latecomer alien to our evolved nature, a tissue-graft that was never going to take. But many things are difficult because they are difficult.

We won't get anywhere by thinking that everything is bound to go contrary with us because our brains are simply not wired to cope with anything post-savannah, that we are naturally lone creatures who have been forced – pretty much against our will – into the society of others, conscripted into the modern world. We won't get anywhere thinking like that because it simply isn't true. New people, new things stimulate us. 'For there is in the mind of man', as Darwin wrote, 'a strong love for slight changes in all things.'

Stimulated by new finds, new technologies and new places, we develop and twist them in a thousand improvised ways. As we do so, the things and places and niches we make change us in a thousand ways too. Only a perfect stranger to human experience would fail to spot how we relish the challenge of this new seed or that new nut to crack. In fact, if you want to describe the incessant probing of the world by the most insatiably curious and dazzlingly resourceful creature that ever walked the face of the earth you would be hard pushed to find a worse verb than 'cope'.

The idea that we struggle to cope comes from a prior commitment to the wiring metaphor, which is itself a piece of bad debt inherited from the computational theory of the mind. The compound interest we pay on this bad debt is moral hazard. To say that our brains are not wired to cope with the way we live now encourages a victim culture. If modern society is rigged against you no-one can blame you for lashing out. In fact if you are coping, then you must be dead from the neck-up. Go wild or go postal – at least that means you're still alive, untamed, real.

I'd say this story – in one form or another – has pretty thoroughly percolated its way through to just about everyone, hasn't it? Not that most people accept it, just that we are all wearily familiar with this line of thinking, with this story outline, this pitch for Human Narrative Number One.

Its thorough percolation is because the Myth of the Stone Age Brain has less to do with East Africa than the Wild West. It is is a backdating of the Myth of Rugged Individualism, a nostalgia for an imagined American frontier, for a home on the range. Its roots are in a philosophy that sprang onto the world's stage at the end of the nineteenth century with the closing of the American frontier.

In 1893 at the Chicago World's Fair, the eminent historian Frederick Jackson Turner delivered a paper for the American Historical Society called 'The Significance of the Frontier in American Life'. No paper ever enjoyed greater influence on American intellectual life.

The frontier, said Frederick Jackson Turner, forged the American character. So long as the frontier was open, rugged individualistic pioneers pitted themselves against treacherous terrain, waged war on hostile tribes, foraged for food on barren hillsides, drove wagons across white water rapids, and faced down bears and wolves. The problem was that the frontier made American manhood too well, and before you knew it the West Was Won, Native Americans were all in reservations, the United States spanned a continent and the frontier was closed. This success would, Turner warned, be the undoing of the national character. Now came the debilitation and degeneration of the bloodstock into fey, weedy, weak-chinned, limp-wristed, whey-faced frivolity. Soon every American male would dwindle to a Bostonian fop in spats and pomade. After all, the Southern-Pacific Railroad were offering tickets for 'The Great Pleasure Route of the Pacific Coast'. The Wild West was now so tame that you could watch it go by from the viewing car of a train.

'The Significance of the Frontier in American Life' spawned the cult of rugged individualism, a cult which still has its adherents in intellectual life, especially in America. The reason horse-riding is not included in a list of newcomers to human experience for which our brains are not wired to cope is because they are part of the stage props for the Wild West. Pinker's litany of things our brains are not wired to cope with is really just a list of things that cowboys can't be doing with. For anonymous crowds read Dodge City, Santa Fe and city slickers in general. For government, police and courts read sheriff and sheriff's deputy. Forget modern medicine, I'll bandage the gunshot wound with my own bandana. I'll suck out the rattlesnake's venom my own damn self. Cowboys can cope with horses, campfires and beef jerky, but they sure can't cope with filling in forms. In the Western *Lonely Are The Brave,* Kirk Douglas plays the last cowboy, unable to cope with modern bureaucracy (written language). In the film's closing frames he and the horse he rode in on are run over by a truck-load of toilets – another newcomer to human experience! – and both die before an ambulance arrives.

If Pinker had in mind East African savannah and not Western prairie then his list wouldn't be your standard modern bugbears. Instead it would include ALL post-Pleistocene newcomers to human experience. If 'anonymous crowds', then why not trousers? If 'written language', then why not, pottery?

The Great Trouser Craze

Would anyone ever say that our brains are not wired to cope with trousers? Yet trousers are a very late newcomer to human experience. The earliest evidence for trousers dates from less than 4,000 years ago. If our brains aren't wired to cope with 'brand-new agricultural civilization', they are definitely not wired to cope with this sartorial parvenu. So how do we explain the Great Trouser Craze which broke out about 4,000 years ago?

We explain it by way of horses. It seems the first trousers were a pair of woollen jodhpurs worn by nomadic pastoral horsebacked herders in Turfan, east Central Asia. If Turfan was Toledo, we might even call these horse-backed herders cowboys.

The link between horse-riding and trousers was established by radiocarbon-dating the pair of woollen jodhpurs found in a Central Asian tomb, in Turfan. The Turfan Trousers are the first evidence we have of trousers. The fact that they were found alongside equestrian grave goods, leads Ulricke Beck and her colleagues to conclude that: 'the invention of bifurcated lower body garments is related to the new epoch of horseback riding, mounted warfare and greater mobility.'*

Horse-riding came late to us but the more we did it, the more it became us. Riding horses stimulated our creativity in hundreds of way. Not least of which was the invention of trousers at Turfan.

Some argue, however, that these are not the first bifurcated lower body garments. Instead they contend that Ötzi the Iceman's goat-hide leggings predate the Turfan trouser.

In the late Neolithic, Ötzi the Iceman fell into the snow and ice of a Tyrolean glacier after being shot from range by an archer. For the next 5,300 years the glacier preserved his clothes until they were found, in 1991, as uncannily intact as his piercing blue eyes. But the leggings Ötzi is wearing are not, strictly speaking, trousers but thigh-high crotchless leather hold ups. This has led archaeologists to conclude that he was murdered while on his way to a fetish party. Maybe his fetish was being shot at with a bow and arrow. Maybe the thrill was in how close the arrows came, the delicate caress of the feathered flights stroking his shoulder, the low whistle in his ear as a shaft whizzed past. Maybe he strayed too far from the archer for his safe word to be heard.

---

* U. Beck *et al.*, 'The invention of trousers and its likely affiliation with horseback riding and mobility: A case study of late 2nd millennium BC finds from Turfan in eastern Central Asia', *Quarternary International*, 2014.

Ötzi lived and died a thousand years before the Turfan horse-men, but it is they not he who wear the trousers. Unlike Ötzi's leggings and loincloth combo, the Turfan trousers have seat, crotch, gusset, and a waistcord that ties at the hip. This makes them the first bona fide trousers. And rather stylish they are too.

Hours of intense work went into the weaving of these trousers. Dark brown in the legs, the seat is cut from a lighter mustard-coloured cloth. Mazey geometric designs decorate the knees. Zigs-zags are woven around the hem and in a horizontal band across the shin. The crotch has been patched with a woven fabric decorated with the same zig-zag motif as the hem.

What led us to trousers was the horse. One newly acquired skill – horse-riding – led to a new piece of kit – the trouser!

Dhoti and shalwar, trunk hose and drainpipes, hot pants and bell bottoms – we took the idea of trousers and ran with it. And though our brains may not be wired to cope with trousers, we seem to do pretty well. Almost always. There are exceptions of course.

One such was the late great comedian Frankie Howerd, many of whose onstage performances were marred by trousers that had uncomfortably ridden up. His brain was not wired to cope with trousers. As he grew older, the condition worsened. In live performance the ratio of actual spoken standup to dealing with acute trouser discomfort tilted calamitously to the point that he sometimes seemed to do nothing but adjust his trousers for hours on end, only managing to gasp out a few words in between. There were few more moving sensations that an audience willing Frankie Howerd to triumph over his trouser difficulties. Then there's the Fatal Trouser Theory of Easter Island …

In 1868 the ship surgeon of HMS *Topaz* noted that 'we found our trowsers ... coveted' by the Easter Islanders, who bought and bartered all the trousers that were to be had. The next English ship to put in found no islanders at all. The theory goes that an epidemic of typhus, spread by English sailors' trousers, may have done for the Easter Islanders. If so, this is the only documented instance of a whole population felled by trousers.

The Fatal Trouser Theory, I should add, lags a very distant second to the ecocide theory, which says the Easter Islanders died out because they destroyed their natural habitat, including cutting all the trees down the better to topple the statues of rival factions. But, even in the unlikely event that the Fatal Trouser Theory should prove true, I submit that the Easter Island extinction is the fault not of trousers but of typhus.

The Xianrendong Potters

'The hunter-gatherer brain was as plastic as our own', argues psychologist Norman Doidge in *The Brain That Changes Itself*:

> and it was not 'stuck' in the Pleistocene at all but rather was able to reorganize its structure and functions in order to respond to changing conditions. In fact, it was that ability to modify itself that enabled us to emerge from the Pleistocene.

Doidge supports this argument with neuroscientific evidence demonstrating the brain's superplasticity, and makes a convincing case that this evidence has short-circuited the 'hard-wired' metaphor. I'd like to make the same case, if it's all the same to you, with reference not to brain-imaging but to clay pots.

About 20,000 years ago, a band of Chinese glacial foragers hunkered down in the Xianrendong caves in Jangxi province to make the first pots. Since this is 10,000 years before rice cultivation in that region, pottery counts as a *hunter-gatherer innovation*.

There is however, an interesting wrinkle to the story of the Xianrendong potters. Just as suddenly as they start making pots they stop. Why? Did they find the evening classes too time-consuming? Too full of people who weren't really there to learn pottery at all but who were just lonely and always going on about their ex-husband? 'Oh, I see you're making a jug. If my ex-husband was to look at that jug, do you know what he'd say? He'd say what a shit jug! Yeah. That's the sort of man he was.'

Or perhaps their pots, jugs and vases just came out wonky, and the handles kept snapping off, and the Xianrendong potters found it hard to cope with failure. If so, that may explain why they appear to have smashed their pottery into tiny pieces, and stomped them into the mud of the cave.

The next pots turn up in Torihama, Western Honshu, Japan and the Amur Basin in Russia, both about the same time as each other. It may be that people in these valleys started making pots independently, or it may be that they bought these pots as a job lot off the demoralised Xianrendongians, who sold them with dire warnings of how pottery is something our brains aren't wired to cope with because it always goes wrong.

Early evidence of pottery shows that we were never as fixed and determined as the wiring metaphor would have us. If Stone Age behaviour was set in stone they would never have begun to experiment with clay. You cannot innovate if rigidly hard-wired.

* * *

Who in the nineteenth century would ever have believed that, out of all their contemporaries, the one whose views would be most in tune with the dominant twenty-first century take on human nature (Pinker's) would be Mrs Gummidge?

'I know what I am', she sniffles in *David Copperfield*. 'I know that I'm a lone, lorn creetur, and not only that everythink goes contrairy with me, but that I go contrairy with everybody. Yes, yes, I feel it more than other people do, and I show it more. It's my misfortun'.'

All this is brought on by Mrs Gummidge being told that Daniel has been drinking in a pub called The Willing Mind. That single pub sign gets closer to the essence of how the mind works than the Myth of the Stone Age Brain ever does.

That we have spent most of our evolutionary history as hunter-gatherers is a truism. The question is what do you do with that truism? Yes, you can use it to say that governments and schools are really heavy, man, like totally oppressive, and anonymous crowds are bumming everybody out. But if we sidestep the melodrama of a stranded Stone Age Man wincing in the bright lights and bustle of the Berlin Kurfürstendamm, or covering his ears against the noise of Liverpool Lime Street, there are better uses to which we can put this fact of pre-history.

Our long backstory can be used to explain some of the things which today are squeezing our skulls. For example, our brains are definitely not wired to cope with the vast deposits of metal particles that have recently been discovered in our heads. It used to be thought that airborne pollution was more a heart and lungs problem than a brain problem. Dissolved in the blood, particulates do their damage in the body, but above the neck, they are checked by the blood-brain barrier, which prevents blood impurities from entering the brain by means of its double row of endothelial and glial cells which link arms and form a blockade. Or so it was thought. It has recently been discovered that ultrafine particles can go straight up the nose to enter the brain via axons in the olfactory bulb. They thought this might explain why, when handling freeze-dried slices of brain tissue, they heard them rattling like tin foil. In Taiwan, Jung, Lin, and Hwang investigated the link between newly-diagnosed Alzheimer's Disease (AD) and long-term exposure to pollution from nanoparticles called PM2.5, which stands for particulate matter with a diameter less than 2.5 microns (a micron is one thousandth of a millimetre). They found a 138 per cent increase in Alzheimer's diagnoses per increase of 4.34 micrograms per cubic metre of pollution nanoparticles.[*] In 2016, when Barbara Maher and her team examined brain-tissue from the Manchester Brain Bank and Mexico City morgues, they found remarkably high concentrations of externally-sourced magnetite pollution particles in the brains of those who had died of dementia and cerebrovascular disease.[†]

The fact that we evolved to be a species dependent on other people and the natural world helps explain some of the things that go wrong when we are separated from them. One problem

---

[*] Chau-Ren Jung, Yu-Ting Lin & Bing-Fang Hwang, 'Ozone, particulate matter, and newly diagnosed Alzheimer's disease: a population-based cohort study in Taiwan', *Journal of Alzheimer's Disease*, 2015.

[†] Barbara Maher *et al.*, 'Magnetite pollution particles in the human brain', *PNAS*, 2016.

that really does seem to be new is the reported rise in otherwise perfectly healthy children who struggle to make sense of facial expressions.

In 2015, psychologist and split-brain expert Iain McGilchrist said that more and more children find it difficult to understand the emotional meaning of facial expressions. He blames this new inability to read faces on children spending too much time on tablets and smartphones. Others blame the parents for not getting down on their hands and knees and spending quality face time with their children, pulling funny faces and barking like a dog. Our playgrounds are full of smartphone orphans. On the rare occasions I see dads taking their children to a playground, I am amazed how little time they spend interacting with their children, and how much time they spend looking at their phones. Didn't they do a lot of texting while they were at the wheel of their car, and nearly ran us over at the zebra crossing? Can't they put it away now that nothing exciting is going to happen like perpetrating a hit and run incident? I hear the heartbreaking sing-song of their children saying: 'Daaad, push meeee, Daaad.' I never understand why children are so patient with parents texting instead of playing. The one thing they should really throw a shit fit about is the one thing they never do. Maybe they know when they're beaten. But the dads, when they reply, on the third or fourth time of asking, never even look up from the screen, they just mumble to their children: 'In a minute. Daddy's just sending this.' When the phone finally gets tucked away, the kid gets a ten-second push, and then it's 'Now remember I've got to drop you off at your play date in five minutes. Off we go. Now don't moan, I said this would only be a short one.'

And that is just those parents lucky enough to have the choice to spend time with their children, squander it as they may. Marilynne Robinson spotlights Welfare to Work schemes that have hustled parents of tiny children out of the home and into the workplace:

> The family as we have known it in the West in the last few generations was snatched out of the fire of economics, and we, for no good reason I can see, have decided to throw it back in again… When we take the most conscientious welfare mothers out of their homes and neighbourhoods with our work programs, we put them in jobs that do not pay well enough to let them provide good care for their children. This seems to me neither wise nor economical.*

… Nor scientific.

Welfare to Work flies in the face of a broad array of neuroscientific and psychological findings about the importance of child–parent face-to-face contact, especially in the early years, the critical period of 'exuberant synaptogenesis'. Employers who benefit from denying children their synaptogenic rights should have to pay a levy for the downstream social costs in which they have invested. Overtime is creating a cognitive disaster.

---

* Marilynne Robinson, *The Death of Adam: Essays on Modern Thought*, 1998.

14. 'FOUND: THE BRAIN'S CENTRE OF WISDOM'*

One day medical interventions might make us wiser. Thomas W. Meeks and Dilip V. Jeste, professors of psychiatry and neuro-science at the University of California, San Diego, made this astounding claim in their paper 'Neurobiology of Wisdom':

> Understanding the neurobiology of wisdom may have consid-erable clinical significance. For example, knowledge of the underlying mechanisms could potentially lead to development of preventive, therapeutic and rehabilitative interventions for enhancing wisdom.[†]

Wisdom is not a biochemical property that you can isolate from a more complex compound. It is not the name of the residue you have left when you boil away silliness. You may learn wisdom after spending £30,000 on a surgical wisdom implant only to come out speaking a bit like Yoda – 'much wiser now am I' – but

---

* Headline of *Sunday Times* interview with Thomas Meeks and Dilip Jeste, 5 April 2009.
† Thomas Meeks and Dilip Jeste, 'Neurobiology of Wisdom', *Archive of General Psychiatry*, 2009.

that's not what Meeks and Jeste mean by developing interventions to enhance wisdom. The idea that wisdom is a substance, that it exists inside us like enzymes or platelets, is a category error. Meeks and Jeste talk about wisdom like it's Sanatogen multivitamins or an effective underarm deodorant: 'Wisdom is considered an important contributor to successful personal and social functioning.'

The wisdom that guided Martin Luther King Jr's campaign of civil disobedience also got him assassinated in Memphis. Being shot dead is considered a major obstacle to successful personal and social functioning. Hermits and dervishes live in the rocks because their path to wisdom demands renouncing successful personal and social functioning altogether. So if wisdom is not like a grooming product or jar of vitamins, how else might it be defined? Meeks and Jeste have a stab at a definition:

A standard philosophical (in Greek, philos-sophia = lover of wisdom) definition of wisdom pertains to the judicious application of knowledge, and most religions have considered it a virtue.

This is a standard dictionary definition (Webster's) so I don't know why they call it a philosophical one. And while I like Jeste and Meeks' judicious qualifier 'most' in the phrase 'most religions have considered it a virtue,' I do wish they'd tell us, which religions have considered wisdom not worth bothering with. The *Apocryphon of John* comes nearest:

On that day, the disciples spake unto Him:
'Master, teach us the path that leadeth to the unveiling of sacred truths and profound wisdom.'

And lo! the Master did say, 'Wisdom? It's not worth the agro.
You should get out more. Have a laugh. Do something
stupid for a change.'
But Master,' sayeth the disciples, 'it is written that by wisdom
come we unto profound acceptance of life as it really is.'
Whereupon the Master did command his disciples to cease
with their bunny rabbit as it was doing his head in.'

What would a wisdom intervention look like? Let's suppose
we discover into which precise synaptic cleft to inject the
inhibitory neurotransmitter glutamate, and it makes us less
prone to rush headlong into situations we do not fully understand, better able to compromise and temporise, to watch and
wait – how will that help us in those situations when those
urging compromise simply haven't understood what's going
on? Who was wisest in May 1940, Halifax or Churchill? If,
instead of drinking Prunier brandy Churchill had been receiving nasal inhalations of inhibitory glutamate, we'd all be
speaking German.

We teach children the virtue of compromise, but we also
teach them that there are things about which there can be no
compromise. Compromise is by no means always wise, especially
not when a point of principle is at stake. For decades, Aung
San Su-Kyi rightly knew that until she had secured Burmese
democracy, she could not budge an inch from implacable
non-cooperation with the Myanmar junta. Fatal to all hope of
democracy would be to accept a job in the junta's agriculture and
fisheries ministry, say, in the hope of rising to a cabinet position, and by patient compromise and careful behind-the-scenes
diplomacy, influencing the direction of policy towards a more

democratic orientation. She knew that that sort of trimming would be the end of everything she was trying to do.

You cannot distil wisdom. What is wise one instant is foolish the next. Both before and after the American Civil War, and for large chunks during it, Ulysses Grant is dim-witted and confused, drinks too much, fails at one business venture after another, slides into despair and despondency. But by 1863 he is almost alone in being able to see one simple truth that escapes more savvy, sophisticated military and political minds. Unconditional surrender. Now, the shape of Grant's brain hasn't changed to fit the times. The times have changed to fit his mind.

* * *

'Remember always that a wise man walks with head bowed, humble like the dust', Shaolin priest Master Kan tells Kwai Chang Caine in the TV series *Kung Fu*. The reason why it's wise to walk with your head bowed in nineteenth-century China is a question of power. In a feudal autocracy you'd best not meet the glare of the Emperor's guards who will kill you as soon as look at you. It is wise for Kwai Chang Caine, and all those without power in the world, to learn tolerance, self-regulation, and every other self-abnegation going, but when we look at the successful and powerful people in the world do we find they walk with their head bowed, humble like dust? If not, then how did they achieve such 'successful personal and social functioning' without being wise? This has to do with power. A failure to factor in power relations can make discussion of wisdom in the abstract a bit waffly:

Tolerance of other persons' or cultures' value systems is often considered an important subcomponent of wisdom.

A perk of power is that you no longer have to be tolerant of persons' or cultures' value systems. A servant is wise to learn what Jeste and Meeks call 'emotional homeostasis' (not blowing your stack) but the Lord of the Manor can take it or leave it.

* * *

What's wrong with 'Neurobiology of Wisdom' is not a question of whether it is premature to claim that neuroscience can locate wisdom. No matter how sophisticated brain-imaging technology becomes in the future, the type of explanation that Jeste and Meeks offer here will never do. It will never do because their concept of wisdom is every bit as innocent of real, lived human experience as the claim we examined in the AI chapter that logic is how we come up with our ethical choices. What is wrong here is the same as what was wrong there, as the next quote makes clear (or as clear as such turgid prose can be):

> The lateral prefrontal cortex facilitates calculated, reason-based decision making, whereas the medial prefrontal cortex is implicated in emotional valence and prosocial attitudes/behaviours.

Calculated, reason-based decision-making is not calculated to be wise. No giddy whim compels oil executives to drill for oil in the Arctic. Psychotherapists grow tired of reminding clients how our self-destructive ideas tend to come disguised as friends. 'No, you're wrong', we reply. 'This time it's different. This time I'm making my decision after careful consideration. I'm going in with my eyes open. I've chosen a sane and thoughtful partner, and her ACAB forehead tattoo is neither here nor there.'

In *A Treatise of Human Nature* David Hume (1711–1776) argues that we all have our reasons for the most unreasonable things you can possibly imagine doing:

> 'Tis not contrary to reason to prefer the destruction of the whole world to the scratching of my finger. 'Tis not contrary to reason for me to chuse my total ruin, to prevent the least uneasiness of an Indian or person wholly unknown to me.

I never understood what Hume meant by preferring the destruction of the world to a finger scratch until I listened to the brilliant Josh Ritter song 'The Temptation of Adam'.

Alone together in a nuclear missile silo 'three hundred feet under the ground', co-workers Adam and Marie, fall deeply in love. As their tour of duty comes to an end Adam fears their love won't work out on the surface. The dark thought comes that if World War III broke out they could stay just as they are, together forever, snug in their bunker with a lifetime's supply of rations. Then Adam has an even darker thought. Nuclear weapons operator that he is, in possession of the nuclear codes, he has to hand an excellent way of starting WWIII unilaterally. The song ends with the line:

> I think about that great big button, and I'm tempted.

'The Temptation of Adam' reminds us that you can reason and calculate yourself into the world's worst decisions. We methodically ponder and strategize our way to the everlasting bonfire, while nodding approvingly to ourselves at each judicious link in our chain of reasoning. In sharp contrast, at the very moment we take a decision that will turn out excellently, we are often beset by voices in our head warning that we have surrendered to our worst self-destructive urges. There are sound reasons for this internal heckler to pop up at moments when we are about to go our own way. We have evolved to care a lot what others think. The internal heckler is a repository of all the loving and sound advice we have absorbed over many years and which we are about to ignore for reasons we can't quite put into words just yet. We are going our own way, but because we set so much stock by their counsel, it leaves us feeling out on a limb.

An interesting case in point is direct action. When people take direct action, they sometimes have to surmount all kinds of internal hecklers. You have always been careful crossing roads. But now, all of a sudden, you and affinity group are sitting down in the middle of five lanes of traffic.

This is why part of direct action training is designed to help you get over the voice in your head telling you that what you are doing is wrong. Before civil rights activist Rosa Parks sat in the whites-only section of the bus in Montgomery, Alabama, and before Ploughshares women broke into a British Aerospace hangar in Lancashire to disarm a Hawk attack aircraft, it was necessary for them to undertake a course of intensive direct action training, in part designed to help them silence these internalised voices, which Jeste and Meeks would like to call 'the medial prefrontal cortex'.

> [T]he prefrontal cortex figures prominently in several wisdom sub-components (eg. emotional regulation, decision making, value relativism) primarily via top-down regulation of the limbic and striatal regions.

The civil service is big on 'top-down regulation' but why should the pre-frontal cortex prefer this management approach? You could argue that this is just a metaphor – that Jeste and Meekes are merely trying to convey how a 'higher' part of the brain damps down urges from a 'lower' part. But a metaphor is never just a metaphor, and this one allows them to personify a part of the brain, and so have it do things in the same way that we would do them. This is, I think, an example of what philosopher Ray Tallis calls the 'smuggling in of consciousness by way of anthropomorphism'. Ray Tallis has been a customs and excise vigilante kicking down the door of one consciousness-smuggler after another. The smugglers pretend to have a physical explanation for everything, but they simply don't. Explanations that pretend to be purely material, have actually smuggled in some sort of conscious agent, and the way they do this is by anthropomorphising the brain. We are all supposed to look down on mind–body dualism, but this is less impressive if we are dressed in the furry onesie of anthropomorphism.

While appearing only to be describing neurochemical reactions, Jeste and Meeks also imply that these brain regions know what they are doing, as if a little homunculus was at work. Since

the pre-frontal cortex is all but synonymous with consciousness anyway, then Jeste and Meeks can imply that it sort of *is*
consciousness, which makes it seem not very far-fetched for it
to be in favour of a top-down regulation.

My point is not that noble faculties of wisdom and love are
sacred temples to which we must never admit the cold light
of science. I am saying that Meeks and Jeste make a category
error in assuming that wisdom can be reduced to particles or
atoms. This fallacy is widespread, and understandable. People
are anxious that if they cannot account for everything in the
language of microphysics then they have surrendered to meta-
physics. Bertrand Russell, that least mystical of men, made the
wise point that just as there are laws of physics, so there are
laws of psychology. Laws that apply to one domain, do not apply
to the other. This has been forgotten. But perhaps that most
mystical of men William Blake can help us remember a plain
and simple matter of fact of life which is that concepts, such as
wisdom, aren't material things:

What is the price of Experience? Do men buy it for a song?
Or Wisdom for a dance in the street? No it is bought with
the price
Of all that a man hath, his house his wife his children.
Wisdom is sold in the desolate market where none come to buy.'

15. THE DEATH OF ALLEGORY

In his brilliant and beautiful poem 'The Death of Allegory', Billy Collins wonders what became of all those personifications of metaphysical concepts found in medieval woodcuts and eighteenth-century statuary:

> Truth cantering on a powerful horse,
> Chastity, eyes downcast, fluttering with veils …
> Reason with her crown and Constancy alert behind a helm'.

Once upon a time, these allegorical figures spent their days travelling between the Valley of Humiliation and the Sunlit Uplands of Joy. But where are they now? In the poem, Billy Collins imagines that they have retired to 'a Florida for tropes'.

But neuroscience has since brought allegorical figures out of retirement, and relocated them to different postal districts of the brain, where Truth, Wisdom, Villainy, Schadenfreude and Envy have each been given plots of land – although Envy's isn't as well situated as the others. (Much to Schadenfreude's delight. A delight that Villainy took no time in telling Envy about, you can be sure. Truth was going to mention it as well, but Wisdom advised against it.)

It took state-of-the-art high-tech fMRI machines to bring back the medieval idea of anthropomorphising abstract concepts. But there is one crucial difference between allegorist and brain-mapper. Neither Hieronymus Bosch nor John Bunyan believed you could actually meet a concept in the flesh. They never confused parable with history, nor revelation for the view from a window. They never believed that the Valley of Humiliation or *Garden of Earthly Delights* were actual places in the physical world. They imagined them as being lots of places and none. While writing *Pilgrim's Progress*, Bunyan did not believe that there is a flesh and blood Mr Despondency who has a daughter called Much-Afraid. But unlike Bosch and Bunyan, the brain-mappers really do believe in the flesh-and-blood existence of abstract concepts, albeit microscopic flesh and blood.

And here's a curious thing. For some reason, they think that oxytocin is the trail scent of abstract concepts. The brain scanners' hounds follow the scent of oxytocin, tracking the Big Foot of Guilt down to its lair in The Cave of Shame, which lies in the folds of the temporal sulcus, or under the shadowy overhang of the posterior angulate gyrus. The overpowering scent of oxytocin has led to a rash of peer-reviewed papers with titles that sound like they might have come from the casebook of Elizabethan alchemists such as Simon Forman or John Dee:

'Oxytocin increases trust in humans'*

---

* M. Kosfeld, M. Heinrichs *et al.*, *Nature*, 2005.

'Intranasal administration of oxytocin increases envy and
   schadenfreude (gloating)'[*]
'Oxytocin improves "mind-reading" in humans'[†]

But change is in the air. Peer-reviewed papers increasingly
address the bad odour to which these berserk extrapolations
have subjected neuroscience. In the decade or so since the first of
these oxytocin papers was published there has been a move away
from reductivism. A good example of this is Meghan Puglia
et al.'s paper 'Epigenetic modification of the oxytocin receptor
gene influences the perception of anger and fear in the human
brain'.[‡] This paper argues that the role of the oxytocin molecule
needs to be understood as part of a dynamic, complex interplay
between brain, body and outside world:

> Phylogenetically, it is improbable that a biochemical with
> such wide-ranging targets evolved to have a highly focused
> effect on specialized processes, like trust or envy. Rather,
> oxytocin likely has a more general effect on basic biological
> systems that ultimately support these complex social-cogni-
> tive constructs.

It is with a sigh of relief that you move away from the Simon
Forman/John Dee quackery in which a single shot of magic
potion (oxytocin) produces jealousy or envy or bitterness, and

---

[*]   S. G. Shamay-Tsoory *et al*, *Biological Psychiatry,* 2009.

[†]   G. Domes *et al.*, *Biological Psychiatry*, 2007.

[‡]   Meghan Puglia *et al.*, *PNAS*, 2015.

into something which at least approaches a recognisably human scale of complexity, namely that oxytocin forms only part of a wider array of developmental systems interacting with each other. This is not to say that complexity is always closer to the truth than simplicity, or that we should respect fiddly theories rather than pithy ones. It's just to say that when it comes to explaining human states of mind a multiplicity of weakly acting causal pathways sounds about right.

Epigenetics has rescued neuroscience from searching for answers in the neuron or the gene alone. The study of non-DNA heredity, epigenetics has entailed a move away from simple linear explanations to ones that involve the integration of multiple causes – a move, that is, from mythology to science.

The enriched conditions of a pregnant rat's immediate habitat, for example, accelerate the *in utero* maturation of her pup's visual cortex, as if daughter cells were being told the outside world is worth looking at. As a result of this epigenetic cascade her pup will be born with precocious eyesight.

The cytoplasm of the rat's brain cells influence how the chromosome modifies gene expression. Within each cell's nucleus chromosomes control gene expression by way of transcription factors such as the enzyme methyltransferase, which mutes *now this, now that* strip of DNA. Once a gene has been methylated its expression is silenced. By the same token, sections of DNA that have lain dormant for generations can be activated by these protein methyl markers.

Cellular cytoplasm, in turn, depends on ecological conditions in the womb, while womb ecology depends, in its turn, on environment, nutrition, social conditions, experience. You may noticed the strange absence of the womb from accounts of evolutionary biology. Unlike genes that code for behaviour, it is not scientifically respectable to talk about the womb, even though genes that code one-to-one for behaviour do not actually exist, whereas wombs do. Wombs, we are told, are about development not evolution. But Eva Jablonka and Marion Lamb persuasively argue that once the zygote has been formed in the womb, the evolutionary story is only just beginning.* What the mother passes through her placenta to the amniotic sac, for example, has been shown to influence her offspring's offspring's offspring. The same is true of her milk. Disease resistance and immune response are major factors in evolution, and a mother's milk passes on a cocktail of immune-boosters, which will influence which of her offspring survive to fertility. And among this cocktail of course is our old friend oxytocin, which brings us back to epigenetic factors on brain development.

Epigenetics teaches us to be sceptical about seeing oxytocin as magic bullet, to be chary of drug companies trying to sell the NHS on plans to use intranasal oxytocin as way of treating behavioural and affective disorders. The weakness of this techno-fix is that it neglects the patient's history. If past conditions have led transcription factors to suppress the oxytocin receptor gene OXTR, then a snort of oxytocin will be useless. Whether or not oxytocin can help us depends on the OXTR gene, and the functioning of OXTR depends on our past. Epigenetics has reintroduced history to the neuroscience, which is what is signally absent from those papers that hunt for allegorical

---

* Eva Jablonka & Marion J. Lamb, *Evolution in Four Dimensions*, 2014.

figures such as Wisdom, Trust, Truth, Virtue and Courage down among the basal ganglia or up on the rocky promontory of the left frontal lobe.

Even though wisdom, anger and fear can never be sufficiently explained by neurobiology, Meghan Puglia and her team show it is possible to conduct useful and scientifically sound experiments into the neurobiology of psychology and emotion. Just as allegory gave way to history in art and literature so we can hope that allegory may give way to history in neuroscience.

16. THE BODY'S IN TROUBLE

In this chapter I want to explore the extent to which our bodies are involved in what we tend to think of as strictly cerebral operations, such as problem-solving or perception.

Esther Thelen performed a series of experiments with babies to investigate bodily influence on problem-solving.[*] She began with a twist on a classic experiment done by the twentieth-century Swiss psychologist Jean Piaget, the 'A not B' test.

Jean Piaget would always introduce himself at parties as a 'genetic epistemologist'. He would then stand there expectantly, with a mischievous smile playing about his lips, waiting for someone to ask him, 'What's that?' But sadly no-one ever did. And so, when Jean Piaget died in 1980, he took the secret with him to the grave. From that day to this, no-one has ever found out what 'genetic epistemology' could possibly mean. It's thought to have something to do with not shouting at your children in supermarkets, but beyond that all is guesswork.

---

[*]   Esther Thelen and Linda B. Smith, *A Dynamic Systems Approach to the Development of Cognition and Action*, 1994.

One of the things he did leave us however, is the A not B test, which goes like this. Once you teach a baby to reach out and take a toy from Box A, if the toy is then placed in Box B, the baby will keep reaching for Box A *even if they see the experimenter place the toy in Box B*. Only after they are eight months old, said Piaget, will babies start investigating whether the toy might in fact be in Box B.

But Esther Thelen discovered that babies can pass the A not B test if you attach weights to their arms or make them stand up. Her results show that babies as young as seven months old can, if weighted down enough, successfully find the toy in Box B.

The baby's memory, she concludes, is in the sensorimotor circuit that is involved in reaching for the toy. By strapping weights to the baby's arms she interrupts this sensorimotor circuitry. A change in the reaching dynamic orchestrates a change the brainwaves. Freed from repeating the same gesture as before, the baby can try a different way of searching, one which activates neural pathways outside the runnels of the old, failed action.

Esther Thelen also looked at babies' spontaneous stepping activity. Newborn babies when held upright have a stepping reflex that vanishes at two months when their legs become too chubby and unwieldy. Once gone, the stepping reflex usually won't return until the baby is at least nine or ten months old, and gearing up to start walking. Here again, Thelen was able to tweak the standard developmental chronology by strapping weights to babies and putting them on a treadmill. She placed seven-month-old pre-walking babies on a treadmill, strapped weights to their legs and found that the stepping reflex returns very quickly indeed once the machine is turned on. It even does so, she found, when babies have one foot on each of two parallel treadmills that are going at different speeds.

I was intrigued by Esther Thelen's findings and so I decided to visit her in prison.

I wanted to know whether the ability of seven-month-olds to successfully reach for the toy in Box B, might be explained by other factors than sensorimotor circuitry. An alternative explanation might have to do with what neurologists call the 'stickiness' of left-hemisphere processes. The left hemisphere is given to 'punding', meaninglessly or automatically repeating a task, going round and round in ever-decreasing circles unable to escape the gravitational pull of pointless repetitive activity. As everybody knows, the way to snap out of the left hemisphere's punding tendency is to go for a wander and come back at the task afresh. Whenever we are stuck at a problem we get up, potter about and often return to find that our unconscious has solved it for us. Perhaps Esther Thelen's experiments show that memory is stored in patterns of movement. Anyone who has ever learnt a tune by heart on the piano, will know about muscle memory, the way your hands remember how the tune goes better than your head, and you find your fingers on the right piano keys before you have sent them there. Her work also suggests perhaps that by bringing the body back into the equation, we escape left hemisphere's tendency to fall into an attractor basin of ever-decreasing circles

Where a lot of neuroscience seeks to dash the brains from the body, to brain us, Esther Thelen's valuable work helps put brain back in the body, and shows the body to be an integral part of problem-solving. The same is true of Edoardo Bisiach and Claudio Luzatti's work with hemineglect patients.[*]

In Milan in 1978, Bisiach and Luzatti were examining two patients, a man and a woman, both with left-sided hemineglect as a result of strokes injuring the right parietal lobe. Even though there's no damage to the eyes, left hemineglect patients lose the left side of their visual field. The added wrinkle is that they are *not aware that there is anything missing from their picture of the world.* They take the right half for the whole picture.

Typically there are three tests which neurologists give to patients. If they fail at all three they have what is called chronic hemineglect. When asked to draw a clock-face, patients with left hemineglect draw all the numerals from one to six, but leave seven to twelve completely blank. They don't think anything is missing. As far as they're concerned, they've drawn the whole clock. Ask them to draw a cat, they will draw the right half of the cat and not notice that everything from the left is missing. Asked them listen to the *Today* program, they will find it to be perfectly balanced broadcasting.

Bisiach and Luzatti asked each patient to draw the Piazza del Duomo, Milan's famous cathedral square. Sure enough, only the right half of the cathedral was there but not the left. Slap bang in the middle of the square stands the huge bronze statue of King Vittorio Emanuele II on horseback. The statue faces right. When asked to sketch the statue, everything from the left visual field was missing, such as the back legs of the horse. Instead of leading his troops into the Battle of San Martino, King Vittorio

---

[*]   E. Bisiach & C. Luzzatti, 'Unilateral Neglect of Representational Space', *Cortex*, 1978.

Emanuele now appeared to be arguing with the front half of a pantomime horse. Once again, the patients saw nothing wrong, but this time they did betray a sense that things were a little odd. When questioned about the statue, they responded that it commemorated how Vittorio Emanuele managed to hold together a troubled production of *Aladdin*.

What Bisiach and Luzatti did next was something that no-one had ever thought of doing before. They poured icy cold water into each patient's right ear. Why? Maybe they were bored. Maybe the patients had been annoying them. We will never know. What we do know are the spectacular results. It was a beautiful, magical Oliver Sacks moment. As if they had never suffered right parietal brain damage, the left side of the Piazza del Duomo swam into view. On the clock tower the numbers seven to twelve suddenly appeared, the king's horse sprouted back legs and a tail, and both patients demanded to know why John Humphrys describes oil-industry lobbyists as 'independent think-tanks'.

Remission of the Milanese patients' hemineglect proved only temporary, alas. As soon as Bisiach and Luzatti stopped sluicing the inner ear the left side of the picture dropped out again. Since the restoration of vision lasts only seconds, it cannot be called a cure, but Bisiach and Luzatti's experiment seemed to show that neuroscience had only been seeing half the picture. What else were they missing?

How can the inner ear possibly influence what we *see*? How can cold water do what brain surgery cannot and restore sight, albeit temporarily?

One possible solution to the puzzle of why sluicing the inner ear with cold water results in temporary remission of hemineglect might be because the inner ear is involved in the onset of all movement, all action. Its semicircular canals, each set at right angles to the other, are involved in the initiation of all activity. Every physical response has to be run past the ear before it gets the nod. If the brain evolved primarily to serve movement, then it may perhaps help our understanding of the brain to reverse the usual hierarchy. Instead of seeing the sensorimotor system as subordinate to perception, sensation might be better understood as part of doing: 'two eyes serve a movement.'

Inside the cranium, the sensory and motor areas lie side by side, two adjacent vertical strips like Togo and Benin. Taken together, these neighbouring West African states make up the sensory-motor cortex. Sensory business is supposed to go on in Togo, while motor activity is coordinated in Benin. But the Milan experiment questions this picture of discrete functions in separate cortical locations. In *Pain: The Science of Suffering*, Patrick Wall speculates whether the Milan experiment shows that 'we can sense only those events to which we can make an appropriate motor-sensory response':

> Could it be that we in fact sense objects in terms of what we might do about them? Could it be that we have erected an artificial frontier between a sensory brain and a motor-planning brain which does not in fact exist?

Could it be that there is no real border between Togo and Benin. This idea of a circuit matches Wall's own conclusion, drawn from a lifetime of studying pain, that pain is not the simple receipt of a stimulus but the stage at which the body has got to in reacting to the stimulus.

Over a hundred years earlier, the same thought struck the American philosopher John Dewey. In his paper 'The Reflexive Arc Concept in Psychology' (1896) he takes issue with 'pre-formulated ideas of rigid distinctions between sensations, thoughts and acts'. He argues that the standard concept of reflex arcs – sensation-followed-by-idea-followed-by-movement – is not a true description of how humans and other animals behave. The status of our attention before the stimulus is an important factor in how we respond to the external event. We are primed to hear a sound in a particular way depending on what we are doing at the time: 'If we are reading, if we are hunting, if we are watching in a dark place on a lonely night, if one is performing a chemical experiment'. After a blazing row our hearing is sensitised to each micro-fluctuation in the decibels with which a cup is put on a table. We are finely attuned to the fractional increase in decibels, because the argument has primed nerve cells to assemble in a particular formation and chemical composition. If a cup is set down even a tiny bit louder than usual then it may mean that the other person is still angry and the row may flare up again.

Instead of an arc, says Dewey, we have a sensorimotor coordination. This helps explain why water in the ear brings the Piazza del Duomo back to the eye. Vision is restored because the broken circuit has been temporarily completed. A relay of Christmas tree lights stops working if just one bulb goes. But tap the dead bulb's socket with an electrically conducting pair of pliers, and the Christmas tree lights up again. (Or catches fire. I'm not 100 per cent sure about the details of this experiment.) Anyway, Dewey is saying, I think, that the sensorimotor coordination is like an electrical circuit.

Patrick Wall's idea that 'we sense objects in terms of what we might do about them' links to Edward Tolman's findings that animals are not passive responders to stimuli. In 1948, Edward Tolman's 'On Cognitive Maps in Rats and Men' showed the 'largely active selective character in the rat's building up of his cognitive map'. If a rat receives a mild shock while running a maze, he stops and looks around for its source. What hit me? Where did that shock come from? Rats are seeking to plot meaning on the map they are making. Immediately after receiving a mild electric shock out of the blue, they look up, down and all around for what hit them.

It's like the gag the old silent film stars used to do when sent sprawling by an inanimate object. Knocked down by, say, a loose roof slate, Buster Keaton puts up his dukes on the lookout for who hit him. The gag is funny for lots of reasons. Taking a pugilistic stance towards the inanimate world is funny. And so is Keaton's readiness to believe he is under attack. But perhaps his reaction is also funny because we recognise ourselves in his shadow-boxing. We laugh because we recognise our complete if momentary bewilderment when we have no idea what hit us, the perfect ignorance with which we hunt for meaning and significance in all accidents great and small. Keaton is mimicking how we all react by doing something, anything before we know what to do.

Are there things here that may help us understand why the Piazza del Duomo returned to those two Milanese patients? Does the sudden shock of cold water in the middle ear activate the search for a stimulus? Does vision come back for a moment like an instinctual inner Keaton putting his dukes?

Our brain's activity patterns, argues Louise Barrett in *Beyond The Brain*, are responsive to the significance of a sensation, rather than to the sensation itself. We are guided by meaning. Our senses probe the world for meaning. But what happens down at level of cells and tissue and dendrites and axons to give something significance? A clue as to how this may work was suggested by Walter J. Freeman's concept of a nerve cell assembly.

In his paper 'The Physiology of Perception', Freeman shows how he measured the activity in rats' olfactory bulbs as rats sniffed different smells, and concluded that 'the neurons in the bulb must first be primed to respond strongly to the input'.

This priming comes by way of the rats' personal history with different smells, a history that has lead them to attach significance to the stimulus, by giving it privileged attention.

We tend to think attaching meaning to something is a high-end, conscious, cerebral deliberate act – and it can be that of course. But allocating significance can also happen unconsciously. The senses, said J. J. Gibson, 'can obtain information about objects without the intervention of an intellectual process'. In 1971, John O'Keefe and Jonathan Dostrovsky discovered 'place cells' in the mammalian brain that fire in response to the significance of a particular location, rather than in response to this place or that.* As rats build internal maps they look to allocate meaning to objects and places in their environment.

* * *

Some of the experimental work touched on in this chapter has served to put body and mind back together again, and to remind

---

* John O'Keefe & Jonathan Dostrovsky, 'The hippocampus as a spatial map. Preliminary evidence from unit activity in the freely-moving rat', *Brain Research*, 1971.

us that, in the words of Mary Midgley, we are not two things but one thing.

But how did we ever get to thinking otherwise? Why are we so chary of including the sensorimotor system in our picture of how the mind works? When did we first start thinking of body and mind as being separate?

Dewey blames this tendency on slave-owning Greek aristocrats who downplayed the work-related aspects of things. The Greeks put Contemplation on a pedestal – or rather they got someone else to do the heavy lifting. (Too much like hard work.) Their patrician habits of thought, argues Dewey, have led us to downplay the significance of work-related aspects of the brain such as the sensorimotor system. Under their influence, we have made the brain a detached consciousness module, an executive, as an overseer. If the brain has to deal with the dirty world, let it do so only as some sort of perception and integration module. That this is a dead end can been seen in the way that most brain books quickly degenerate into playing Edwardian parlour games based around optical illusions. Is this a picture of a duck's bill or a rabbit's ears? Two black noses or one white candlestick? And here they will stay until they grasp the nettle that perception is part of doing, that the sensorimotor system is a dynamic circuit, and that the 'primary relation between the mind and the world is interaction not representation'. *

And perhaps this last idea – which Wells calls 'ecological functionalism' – takes us a few steps further towards understanding why the Piazza del Duomo floats into view on a tide of cold

---

* Andrew Wells, *Rethinking Cognitive Computation*, 2006.

water. For a few seconds, the dormant right parietal snaps into life when we think we are interacting, on receipt of a signal from our ear. Instead of the stillness of sleep the brain is frozen in a moment of action, like the statue of the king on his horse.

17. THE FABERGÉ BRAIN

The technicolour brain is one of the great icons of our time. You see the brain icon everywhere. The one on the novelty mug beside me is typical. It has a gold cerebellum, a pink occipital lobe set off by a sapphire medulla oblongata, an emerald parietal lobe, electric-blue pons and purple frontal lobe. This is the human brain as Fabergé egg.

For all the talk about us being hard-wired, the Fabergé brain is strictly wi-fi. Gone are the pale threads of the brain's looping interconnections, the association fibres that criss-cross the white matter. Gone are the flowing cranial nerves, the hairnet of the blood-brain barrier, and the long dreadlock of cerebrospinal fluid that hangs down to the small of the back. To include the projection fibres will mess up any brain diagram, because the projection fibres not only connect different brain regions to each other, but also connect the brain to the spinal cord. To include the projection fibres makes the borders between brain and body seem arbitrary? Where do you draw the line between brain and body? At the spinal nerves? At the visceral nerves? Some argue leave out the 600 million neurons in the gut? Is the gut's enteric nervous system – with its 600 million neurons – a satellite of the brain? Not contiguous but still more brain than brawn? A school of thought even says that the sensory neurons within muscle spindles should be seen as an offshoot of brain, too.

Now, I do understand that the map is not the territory, but I think that the disembodied Fabergé brain is symptomatic of that strange desire to go clear of biology which we have been exploring in this book.

We find the roots of the urge to shuck off nature, and the origin of the idea that mind is independent of body, two-and-half thousand years ago, in a strange commune in Crotone, a numerology cult that found a mystical significance in the octave, strictly prohibited the eating of beans, and who were prepared to murder anyone who revealed a mathematical secret. For here in Crotone, Calabria, on the south coast of Italy, in the sixth century BCE, Pythagoras revealed to those initiated into the higher level of his commune the One Big Truth at the heart of the cosmos: 'All is number'.

Murder in Calabria

In *Pythagoras's Trousers*, Margaret Wertheim says that 'all knowledge was kept secret within the community, and one member was expelled when he revealed the mathematical properties of the dodecahedron.'

He was lucky just to be expelled. Another cult member did not get off so lightly, but then again his crime was greater. The Pythagorean creed was based on the profound rationality of number, if anyone inside the cult revealed to outsiders that numbers were not always rational, then the whole philosophical system was imperiled. When Hipparchus of Taranto leaked the secrets of irrational numbers – pi and the square root of two – he was murdered. Classicists are divided whether Pythagoras drowned Hipparchus with his own hands or ordered other members of the commune to drown him. Either way, the punishment for telling people of the limitation of numbers was death.

For Pythagoras numbers were divine, pure and unsullied unlike the things of the earth. The sensual world tempted the weak away from apprehending the revealed truth of the universe, a truth encoded in number, ratio and proportion. A world of perfect order and harmony. And so, when Hipparchus leaked the ugly facts of incommensurables and irrational numbers, Pythagoras lost his mind.

After the murder Pythagoras realised that in his rage he he had been too hasty. He killed Hipparchus too quickly. If he'd been less incensed, he would have kept Hipparchus alive long enough to confess the names and addresses of everyone to whom he had betrayed the secret of irrational numbers. All the Pythagoreans got from water-boarding Hipparchus was the words 'someone in Taranto', before the they pushed his head under. Now a terrible question hung in the air: who else knew about pi? Who was this someone in Taranto?

Archaeologists recently unearthed a fragment of a play written by Aeschlyus about the aftermath of this murder.

CHORUS: Now, Pythagoras sails alone across the Ionian
Sea to Taranto,
Where down at the dock he finds a gang selling quadratic

equations and differential integers – there are quite a few numbers rackets going on. Before long his suspicions fall on wheelwright called Luciano Bertani.

Scene: Bertani's shop.

PYTHAGORAS: How much for the golden wheel rims you got hanging there?

BERTANI: Oh, they're not really for sale. Just a showpiece.

PYTHAGORAS: But if they were?

BERTANI: You'd be looking at two thousand drachma.

PYTHAGORAS: Well, it so happens I've got two thousand drachma on me right now, and I'm looking for gold rims for my chariot.

BERTANI: This is terrific news. I never thought I'd ever sell the golden wheel rims. How big's the wheel?

PYTHAGORAS: I'm afraid I don't know off-hand.

BERTANI: Never mind, just bring the chariot in tomorrow and we'll check these wheel rims are the right size.

PYTHAGORAS: I'm leaving for Crotone this very afternoon and not due back for another ten years.

BERTANI: Oh no. You sure you haven't got the wheel's circumference jotted down somewhere?

PYTHAGORAS: All I have is this wooden wheel spoke is from one the chariot wheels. Oh, but what am I saying? This is no good to us at all, is it? After all, if we measure the spoke that will only give us the radius – and it's not as if there were any way of calculating the circumference from the radius is there?

BERTANI: I guess not.

PYTHAGORAS: Good day to you.

BERTANI: Wait a second.

PYTHAGORAS: …Yes?

BERTANI: Let me have another look at that spoke. Maybe we can work something out.

PYTHAGORAS: You want a closer look at this spoke? Here it is!

Clubs Bertani with spoke.

CHORUS: Sing, chorus, for news tell we
Of doomed and murdered Bertani,
Oh heavy news, oh mortal blow,
That murder should come to Taranto,
The autopsy report noted a strange detail of this crime
His head was bludgeoned precisely 3.14159 times.

Pythagoras influences Parmenides and then of course Plato. For Plato, the things of the earth are bastardised versions of the Ideal. The Ideal Form of a chair is in heaven, a bastard form of chair is on earth. Plato is like someone who has just lost heavily in a betting shop, all he can see is bastard chair, bastard table, bastard little blue pen.

Plato's Ideal Forms, his repudiation of nature, influences the early Christian church and finds expression in ascetic doctrines such as *contemptus mundi* or contempt for the earth.

'[W]e look not at the things which are seen', wrote Saint Paul, 'but at the things which are not seen: for the things which are seen are temporal, but the things which are not seen are eternal."[*]

Saint Paul was writing to encourage missionaries based in Corinth by reminding them of the impeccably Greek credentials of the new teaching. Yes, they are pioneers in an outpost, pushing the boundaries of Christianity's influence to within fifty miles of Athens, but they should take heart, Paul reminds them, from the fact that the gospel will be familiar fare to the locals. It wasn't as if they had to sell Corinthians on a blue elephant called Ganesh. This was Platonism plus Son of God. Eternal meets temporal. If you can't sell that to the Greeks then you shouldn't be in the evangelism business.

As against this Pythagorean/Pauline *contemptus mundi*, another Christian tradition, no less vigorous, considers the things of the earth to be sacred, endowed with meaning, and given to us by God His wonders to behold. What leads Johnson 'to strike his foot with mighty force against a large stone' is his fury at Bishop Berkeley trying to snatch God's gift of Creation away from him, and deny Johnson the evidence of his own God-given senses. Reason and Sense (the evidence of the senses) are the completion of each other. If Sense can only err, then what – Dr Johnson would like to know – is to stop Reason from freewheeling? One hundred years earlier, the case of René Descartes provided a cautionary tale of just what happens when Reason slips a gear.

---

* 2 Corinthians 4:18, the *King James Version*.

The Man Who Mistook The Entire Human Race For Hats

One day René Descartes looked down from his window at all the hats going by in the street and was troubled by the fact that he couldn't be certain of anything.

'How do I know there's people under those hats', he asked himself. 'What if those hats are being worn by automatons? Clockwork automatons wound up and sent this way to trick me? How do I know it's not all a trick by some Deceiver? What if there is nobody under those clockwork hats at all?'

Inexplicably, people took these questions seriously. Why didn't someone just lay a hand on his shoulder and ask:

'Is everything all right at home, René?'

'No, I can't pay my rent!' he would probably have sobbed.

'Well, look let's see if we can't write a letter to your landlord together, asking if he would consider allowing you to defer payment? *D'accord, René? Ça marche?*'

In this way we could have not only helped René Descartes but also saved subsequent generations from wasting their time trudging over arid philosophical shale. As well as helping him with the rent, we could have offered other practical advice too.

'Tell you what, René, why don't you go down to the street and ask one of those passers-by if they are in fact clockwork. They'll probably punch you in the face, but once your nose stops bleeding we can go for a coffee and a good old natter chat and you can tell me what the trouble is.'

But that, alas, was not the response. From that day to this, philosophers have accorded his panic attack the status of epiphany. Four hundred years have been spent saying:

NEUROPOLIS

He's right you know, they could all be mechanical hats on a conveyor belt. That's torn it! I mean, you just don't know, do you? Descartes has called everything into question, and no mistake. Maybe everything is an illusion.'

Descartes in the abattoir

When not doubting everything, Descartes was a keen anatomist. In Amsterdam, he would go to abattoirs and buy freshly slaughtered carcasses to take home to his dissecting room. When visitors called, he would flourish an arm over the animal cadavers, dismembered limbs and jointed bones, and declare: 'This is my library and these are my books.'

It would have been very tempting to reply: 'But how can you be sure they are books and not animal carcasses? I mean, I'm looking at this now, and I don't see any books, all I see is an ox's tongue and the flayed carcass of a goat. What if it is all an illusion perpetrated by a demon?'

* * *

Things began to go wrong in 1632, when Descartes was at work on a book with the magnificently title *The World*. Imagine the look Descartes gave people who asked him what it was about. Or – better yet, who told him that it didn't sound like their sort of thing. When Descartes heard that the Inquisition had arrested Galileo, he shelved *The World*.

He would no longer claim that we could trust our sense because … that might open him up to accusations that he depended on nature rather than God for his certainty. Instead, Descartes allowed that maybe everything he saw around him was an illusion.*

He didn't just turn away from the world and retreat into his own self. Instead, he repudiated the outside world, and razed the earth, thereby cutting a pattern whose template we have seen stamped all over neuroscience.

Like St Ives, Descartes is famous for his fudge. Descartes' fudge is known as mind–body dualism. 'The mind', he declared, 'can operate independently of the brain'. (Which only goes to show that the pen can operate independently of the brain.) To say that we are made of matter without soul, however, would be impious and so he came up with his fudge: within the material brain the pineal gland houses the immaterial soul. The little pink pineal gland was the intersection of the eternal and temporal. Somewhere inside this hormonal gland was the *ne plus ultra* of the material world, of physiological explanation. The pink pineal was the one part of the mind that operated independently of matter.

'Thus is Man that great and true Amphibium', wrote Sir Thomas Browne, Descartes' contemporary, 'whose Nature is dispos'd to live, not only like other creatures in diverse Elements, but in divided and distinguished worlds.'

The divided and distinguished worlds are the material world and the spiritual realm. Amphibiousness is frowned on nowadays. We are supposed to be able to account for everything materially.

---

* Carl Zimmer, *Soul Made Flesh: The Discovery of the Brain and How it Changed the World*, 2004.

'The brain', says Brian Cox matter-of-factly, 'operates according to the principles of physics.' Setting aside the fact that half of physics doesn't operate according to the laws of the other half, it seems to me that we are indeed amphibians, although not in the sense that Thomas Browne means. Our minds are created by rich interplays of inherited ideas, shared understandings, emotions and experience – none of which physics is adequate to explain. Our nature is disposed to live in the divided and distinguished worlds of culture and biology.

The fact that we cannot locate the sense of loss in the nucleus accumbens of the basal ganglia doesn't mean that our sense of loss in an illusion. If someone feels loss, says Mary Midgley, then that loss is a real thing in the world. In the case of someone suffering grief, the sort of explanation that physics provides is the sort of thing Dr No or Blofeld would say after liquidating your best friend:

> Well, your so-called grief is actually just a substrate of some disrupted electrochemical neurotransmitters. Neurons, which used to be stimulated by the company of your friend, you see, are no longer transferring potassium and sodium ions through the axon's semi-permeable membrane. So what you experience as loss is simply an illusion created by these chemicals. Now all you have to do is plug yourself into an ioniser that will replace exactly that biochemical mix and – lo and behold – your so-called grief will vanish.

This is the part, where Bond says to one of Dr No's henchmen, 'You do realise he's quite mad?'

There are more ways to get to the truth of things than physics. If someone tells you about their loss and your response is to produce from your backpack the tri-axial probe of physics, you have simply produced evidence of your own disqualification for the task of understanding human experience.

Try telling that to Francis Crick, though. In *The Astonishing Hypothesis,* Francis Crick goes for the full Dr No:

> 'You', your joys and your sorrows, your memories and ambitions, your sense of personal identity and free will, are in fact no more than the behaviour of a vast assembly of nerve cells and their associated molecules.

Consciously or unconsciously, Crick is half-remembering a famous line from *The Emotions* (1922), a classic of neuroscience co-authored by Danish anatomist Carl Lange and American philosopher William James:

> We owe all the emotional side of our mental life, our joys and sorrows, our happy and unhappy hours to the vasomotor system.

By echoing not just Carl Lange's general argument but even the particular phrase 'joys and sorrows', Crick disproves exactly what he sets out to prove. Accidentally, but irrefutably, Crick demonstrates that our mental life grows from the rich soil of shared history, culture and learning and the field effect of a community of minds. Crick's allegation that our memories are no more than the behaviour of a vast assembly of nerve cells cannot be true since memories have to be *about* something, and that something can only be found outside the skull, in this case Carl Lange's idea. No-one has ever had any memory of that time when afferents to the caudate nucleus got mistaken for dopamine neurons in the putamen. I suppose the fallback position for hardcore Crickians here is to say: 'Aha, but it's still Crick's molecules talking to Lange's molecules.' But that is a small and airless chat room that we need not visit.

Where Carl Lange uses the first person plural as inclusive testimony to our common humanity, Crick wants only to call other people's identity into question, never his own. (A tic shared by Ramachandran, if you recall.) He is never happier than when putting you in inverted commas: '"You", your joys and sorrows …' There is no record, however, of his ever talking about 'so-called me'. Now, I don't think Crick's use of the second person is solely due to his sense of superiority. I think the usage also helps disguise the melodrama, which jumps out at you as soon as you put Crick's words into the first person:

'Me', my joys and my sorrows, my memories and ambitions, my sense of personal identity and free will, are in fact no more than the behaviour of a vast assembly of nerve cells and their associated molecules.

To which we would naturally respond by asking: 'Is everything all right at home, Francis?' It's no less melodramatic in the second person, but the melodrama is easier to hear in the first, when we are not under attack from sadistic vitriol.

Whether the brain operates according to the principles of physics is more than anyone knows. It should do, but it doesn't seem to. So now what do we do? For Crick the response is to start haranguing everyone for being nothing and no-one. You? Joys and sorrows? Ambitions? Memories? Don't make me laugh. You're nothing, you hear me? Nothing.

'The fanatic,' says George Smiley in *Tinker Tailor Soldier Spy*, 'is always concealing a secret doubt'.

Almost everything worth knowing about our minds remains a mystery. We have got so used to solving mysteries, that we are intolerant of the very concept of the unknown or unknowable. We suspect that ignorance must be wilful obfuscation, a refusal to bow to empirical results, a secret religiosity. But sometimes there simply are mysteries. To pretend that there are none does not dispel mystery, but only replaces reality with a mystical idea of science. As Bertrand Russell pointed out, the zeal for materialist explanation in advance of the facts creates a new mysticism.

18. A STRANGE KIND OF REALISM

There is no reason without emotion, and no emotion without reason. It's a two-way street. We go wrong when our thoughts and our actions are too cold and we go wrong when they are too hot. As Bertrand Russell put it, beliefs create passions as much as passions create beliefs. Emotion and reason are intertwined. In *Kluge: The Haphazard Construction of the Human Mind*, Gary Marcus is having none of it. He deplores what he calls the 'emotional contamination' of human thought:

> Instead of an objective machine for discovering and encoding Truth with a capital T, our human capacity for belief is haphazard, scarred by evolution and contaminated by emotions, moods, desires, goals and simple self-interest ...

Kluge is an extended riff on how humans' brains are clumsy, confused and lumbering, whereas computers are smart, elegant, and well-designed. You'd think Gary Marcus would like people better since according to him we are as predictable as machines, and our actions follow neurobiological laws of human behaviour:

'The more we are threatened, the more we cling to the familiar'.

Well, sometimes we do and sometimes we flee to a foreign country. Then he says:

> Another study has shown that all people tend to become more negative toward minority groups in time of crisis; oddly enough this holds not just for members of the majority, but also for members of the minority groups themselves.

Divide and rule may be a hallowed fantasy of realpolitik strategists but life doesn't always work that way. 'We all charged towards Cable Street', recalls anarchist historian Bill Fishman of the famous 1936 East End battle against fascists:

> At the bottom end, an overturned lorry was used as a barricade and we blocked the road – Hasidic Jews with little beards and great strapping Irish dockers all standing together. People began to throw down their mattresses to block the street ...

Before you can claim that all people under threat hate minorities, you need to account for the people on Cable Street who threw mattresses out of the window, and for the dockers and hasidim standing shoulder to shoulder. Were these people mental defectives, lacking a properly developed ventromedial striatum in the frontal V1 region perhaps?

To illustrate how the wonky shopping trolley of the mind is always veering away from 'Truth with a capital T', *Kluge* celebrates a study in which male and female subjects were asked to read an article on how caffeine endangers women's health. Sure enough:

women who were heavy caffeine drinkers were more likely to doubt the conclusion than were women who were light caffeine drinkers; meanwhile men, who thought they had nothing at stake, exhibited no such effect.

Not one woman in the study can see past her own unconscious bias, which the study calls her 'motivated reasoning'. The coffee is speaking through her and she's too stupid to know it. What she takes for sound judgement correlates to how many cubic millilitres of coffee she drinks per day. Most ignorant of what she's most assured, she has no inkling of the extent to which her views are as pre-determined as the factory settings on a Gaggia coffee machine. This triumphant conclusion of contamination by emotion, however, is not the triumph it thinks it is because it has failed to control for contamination by the outside world, and revealed a rather strong bias of its own.

What if some of the women have learnt that coffee is a good source of antioxidants? That's a fairly well-known fact, so why is it impossible for them to have ever heard it? If a woman knows about caffeine's antioxidant properties then her reasoning can no longer be dismissed as motivated. Now you have to call her reasoning fact-based – unless you are some sort of cad.

Before you can prove that the women are in denial and exhibiting symptoms of 'motivated reasoning' there are three minimal factors you must include in your experimental design:

1. You need to show that no rational reason exists for anyone to doubt the article's claim that drinking more than a few cups of coffee a day is risky for women.

2. You need to make sure that the women are as perfectly ignorant of the science as vervet monkeys in a laboratory, and
have never had previous access to any relevant scientific or
medical information at any point in their lives.

3. And finally, you need to show that self-delusion exhausts
the possible spectrum of psychological reasons for anyone to
speak out when told their ignorance is killing them.

If all three criteria are confirmed – the irrefutability of the article,
the perfect ignorance of the women, the irrationality of their
response – then the experimental results stand.

News stories alternate between saying coffee is harmful one
week and beneficial the next. Is it impossible that a woman may
base her increased uptake of coffee on what she reads? The perfect
ignorance of the women can be ensured only so long as they can
be shown never to have had any contact with the outside world.
Outside the lab, a woman runs the risk of hearing about the
twelve-year long Tokyo National Cancer Centre study which
discovered that women who drink three or more cups of coffee
a day have half the risk of developing colon cancer compared
to women who drink no coffee at all.* On learning about this,
a woman would have good cause to make her espresso a *doppio*.
She would certainly wish to dispute an article that sought to
double her chance of colon cancer. The experiment has failed
to control for an outside world teeming with facts, figures and
information, which may have given the women good reason

---

* Kyung-Jae Lee, Manami Inoue *et al.*, 'Coffee consumption and risk of colorectal cancer in a
population-based cohort of Japanese men and women', *International Journal of Cancer*, 2007.

both for drinking coffee by the jug and for disputing the claim that coffee is bad for them.

You could argue that having heard about the health benefits of tea and coffee doesn't damage the experiment's conclusion in the slightest. After all, the women who abstain from coffee may also have heard about such studies, but they do not dispute the article because they have no self-interest in doing so, since the risks are not ones they run themselves. Therefore the study's hypothesis about motivated reasoning and how we lie to ourselves is as good as confirmed.

Against this, I would say that while the abstemious women may indeed have read the article, the coffee-drinking women are more likely to have remembered it, thanks to the proven way that coffee helps consolidate memories, as demonstrated in peer-reviewed papers such as 'Post-study caffeine administration enhances memory consolidation in humans'.[*] In fact studies show that caffeine may help not just only memory but comprehension, too. In the 1970s Claudio Castellano discovered that caffeine improved the ability of mice swimming in the dark both to work out and to remember the route through a maze. The 'results show that caffeine administration … was followed … by a facilitation of learning and consolidation processes.'[†] This being so, that double-shot flat white or pot of builder's tea may have raised those women's levels of acuity to alert them to mistakes in the article.

---

[*]   D. Borota, E. Murray *et al.*, 'Post-study caffeine administration enhances memory consolidation in humans', *Nature Neuroscience*, 2014.

[†]   Claudio Castellano, 'Effects of caffeine on discrimination learning, consolidation and memory in mice', *Psychopharmacology*, 1976.

What about the criteria number 3, the psychological one? Do denial and self-delusion exhaust the psychological reasons for someone to speak out when told her ignorance is killing her? I think not. There is another rather obvious psychological reason why they might well dispute the article. No one else has been accused of behaviour both self-destructive and ignorant. Not the men and not the decaf women. Since the tea and coffee drinking women have had their judgement questioned, it is only logical that they should rise to their own defence. But, for the author of *Kluge*, to defend yourself from an article saying you are stupid proves your own stupidity, your inability not to be able to see beyond self-interest and bias.

The tea and coffee-loving women, it therefore seems to me, have all the lose-lose options of witches thrown into the river. In the 1600s, the Witchfinder General's notorious double-bind was that if a woman floats she's a witch, and if she drowns she is innocent. It is exactly the same with this coffee experiment. If she agrees with every dot and comma of the article then she is an independent thinker, free of any taint of 'motivated reasoning'. But if she disputes the claim that coffee is risky, she is a zombie contaminated by emotion.

This double-bind is, of course, a well-known contaminant of the way men assess women. You see it with men's attitudes towards women drivers. If a woman drives with such ease and confidence that she hangs one arm out the window, she is accused of drying her nail polish rather than concentrating on the rules of the road. If she sits bolt upright with both hands clasping the wheel, however, she will be told she looks scared, unconfident, like a learner, and that her posture betrays that she secretly knows driving to be beyond her competence. Surely even a child understands that someone who is accused of something will defend herself in a way that someone who hasn't been accused of anything won't. Children also, by the way, understand that the views expressed by characters in a story are not necessarily those of the author. This simple fact totally escapes the author of *Kluge*:

> Are human beings 'noble in reason' and 'infinite in faculty' as William Shakespeare famously wrote? Where Shakespeare imagined infinite reason, I see something else …

When Prince Hamlet of Denmark, an emo student dressed in black, says those words, he is striking a pose as mordant sceptic. You might look at the starry heavens and see a majestic golden firmament but he will have you know that the universe is a miasma reeking of gas and decay. Clearly, he's been reading a pre-Copernican Hawking who has told him the sun is a very average star in the outer suburb of one among a hundred billion galaxies and not a glorious wonder at all. He's been reading brain science books, too, it appears, because if his college pals entertain any good opinion of humans, then Hamlet is going to put them straight on that as well. Humans, he tells them, are 'a quintessence of dust'. A dried up, chemical scum.

Now, Hamlet's nihilism is endearing because his heart's not in it. In fact, Rosencrantz is finding it difficult to keep a straight face. When he hears the prince say that things have been so bad lately that he hasn't even been taken his usual pleasure in men, Rosencrantz goes 'Fnaar, fnaaar', and Hamlet has to tell him off. We know that Hamlet is not the nihilist he pretends to be because he is too pleased to see his college pals, too excited by the traveling players, too affectionate towards Horatio, and altogether too warm. Then there's the fact that what Hamlet says here about human nature won't be the play's last word on the subject. And when that word comes it won't be a word either, but actions involving the stiff-backed leader of a foreign army, a guilty mother, a pile of bodies and a poison cup.

Coincidentally, *Hamlet* is another example of how we are taught to ignore the outside world. We are always told this is a play about procrastination. Or that Hamlet's fatal flaw is that can't make up his mind what he do, which leads to tragedy, because while he shillies and shallies the situation gets out of hand. But is the message of the play: Don't Delay, Kill A King Today? Is Shakespeare saying: Killing A King, Ain't No Thing? I think the play is about a young man who is in a simply impossible situation – just like the coffee-drinking women are in an impossible situation: damned if they do and damned if they don't agree with an article about the dangers of the poison cup.

I've focussed very closely on this coffee-denial experiment because it is a representative one. Brain science books so often seem intent on making us doubt our cognitive abilities, our memory, our vision, our reliability, our eyewitness testimony. Now, I know there's an argument that it is good for us to be reminded of our fallibility. A proper appreciation of human tendentiousness will make us less strident, less self-righteous, and more humble. We will become better listeners and better democrats. But these civic virtues are bought at too high a cost. To be told that everything you think is wrong and merely the product of motivated reasoning and anachronistic genetic response does not help produce active citizens. It makes us passive and full of doubt, our native hue of resolution sicklied o'er with the pale cast of thought. We will hesitate before throwing our mattresses out of our windows, and end up never helping the dockers and hasidim fight against the fascists.

Consider the invisible gorilla at the ball-game experiment. A staple of brain books is to retell the famous experiment about how no-one notices the man in the gorilla suit in the videotape of the basketball match. While they watch a video of a college basketball match, students are told to count how many passes the blue team make. Narrowly focussed on this numerical task, they fail to notice when a man dressed in a gorilla suit walks into the middle of the screen, dances around and then leaves. This experiment is taken to prove once and forever the unreliability of eyewitness reports. But does anyone really believe this works except when looking at a screen? You know and I know that if a man wearing a gorilla suit danced onto the field of play in the last minutes of an important derby – say a Celtic versus Rangers football match – he would be lucky to escape with his life.

Now, on one level I applaud this experiment for showing that our eyes do not, after all, work like film and TV cameras – as Jack Gallant's brain-decoder experiments suggest. I applaud the experiment for confirming Edward Tolman's experimental findings in the 1940s, *viz* the largely active selective character of our response to stimuli. I applaud the experiment for lending support to Louise Barrett's argument that the brain's activity patterns are responsive to the significance of a sensation rather than to the sensation itself. All this is to the good, but none of it is the intended take-home from the invisible gorilla at the ball game experiment. Instead, what we are supposed to take home is that we live in a world of delusion and make-believe that makes us next to useless in a court of law, when called as witnesses.

I have an alternative hypothesis for our inability to see the gorilla, an explanation that is based on evolutionary psychology, and elaborated in a paper I recently submitted to the journal *Science.* Here is a précis of that paper, which is called 'Genetically acquired inability of human retina to perceive gorilla'.

Since gorillas are peaceful unless disturbed, humans evolved not to see gorillas. Inquisitive species that we are, early humans no sooner saw gorillas than they started annoying them and were killed. In this way, nature selected for those of our ancestors best able to ignore gorillas.

Of course, this acquired gorilla blindness carries costs as well as benefits. The fact that we can only see gorillas when we make a conscious effort of will means that sometimes we never see an angry gorilla until it is too late. Other times, unscrupulous people have taken advantage of our acquired inability to see gorillas.

To test my hypothesis, I showed the Zapruder film to a silverback western lowland gorilla called Clio. At Frame 331, Clio began grunting as the Kennedy motorcade rode through Dealey Plaza. I paused the film, and zoomed in. Nothing. But then Clio extended a long arm and very slowly tapped her leathery fingers on the grassy knoll where a man in a gorilla suit aims a Winchester rifle straight at President Kennedy.

I await a response from *Science.*

The case of coffee-drinking women is representative not just of the many neuroscientific and psychological experiments that seek to prove we are more or less delusional. It is also representative of a central theme in this book, estrangement from the world. In this case, the estrangement is reflected in an experimental design that assumes volunteer subjects not to have had much contact with the world outside the lab.

This desire to shut out the world and deny its influence takes many guises, as we have seen, but the really odd thing, which I want to point up here, is that this turning away from reality goes hand in hand with a tough-talking realism. Dehumanising and pessimistic popular accounts of how the brain works are justified by appeals to pragmatic realism. These are the facts, plain and unadorned. This is what actually makes people tick. It may not be a pretty picture, but this is how the mind actually works, like it or lump it. And yet for all the jaded, worldly-weary, cynical pose, brain science writers are remarkably unworldly and innocent about the world.

This innocence can take the form of simply failing to notice that there are many more security guards than there used to be (another type of gorilla blindness). Or it can take the form of a more general indifference to the outside world. 'We mostly walk around in our own mental worlds', says David Eagleman rather tellingly, 'passing strangers in the street without registering any details about them.' But doesn't that rather depend on where you live? You might not last very long going round Naples, Karachi or Hackney with your head up your arse. (Perhaps not noticing the world around you is what makes all those futuristic fantasies seem more plausible than they do to the rest of us.)

We saw this unworldliness in Steven Pinker's ignorance of how people really enjoy large gatherings and anonymous crowds. (At time of writing, people are paying £238 for a Glastonbury ticket before they even know who's playing. They go for the festival itself.) In this chapter the strange innocence of the world came out in a bizarrely purblind notion of how a stage play works, and an innocence of the likelihood that someone accused of ignorance and stupidity is not likely to agree with the accusation.

You have to wonder whether simplistic brain science appeals to those for whom human behaviour is especially puzzling. Not because they especially curious, but because they are more than ordinarily perplexed by why people do what they do. Unable to understand what is going on before their very eyes, perhaps they hope to do better by restricting their investigations to the pitch dark of the cranium.

* * *

To enter the gates of Neuropolis you must first be shorn of your belief in free will and in your ability to make sound decisions. On the approach roads, you will be asked a series of questions every one of which you will fail. But that is as it should be, that is good. For only when you admit what a kluge your brain is ,and that human memory is hopeless, only then will you have proved yourself ready for the pharmacological reboot, for the psychological refit, for new memories in place of old, and a life of service to the algorithm. Look up! Behold the gates swing open wide! Forward we go – to Neuropolis!

19. TOO MUCH MONKEY BUSINESS

What does it mean to say that the brains of men and women are different? Now, just so there's no misunderstanding, I am not one of those people who believe that gender is entirely socially constructed. I do believe in the existence of innate differences between men and women, I just don't think they're as binary as we're told they are.

In my own life, the existence of innate difference between men and women happens to be rather important: I don't have any male friends, not a single one. I used to think that this is because I'm on the same intellectual and emotional wavelength as women, until it was pointed out that I don't have any female friends either.

In *We Are Our Brains*, Dick Swaab states that the hormone prolactin makes women enjoy cleaning more than men, which must be why so many more cleaners are female. An excess of the hormone prolactin in males presumably sends men into becoming roadsweepers or dustmen. That may explain why dustmen collect all those soft toys and dolls that they stick on the radiator grill and why they are so effeminate.

One study very popular with Dick Swaab is a paper called 'Sex differences in response to children's toys in nonhuman primates', first published in the journal *Evolution & Human Behaviour* in 2003.

To prove that toy preferences between boys and girls are innate and not the product of parental pressure and the toy industry, the paper's authors Alexander and Hines took a mixed group of male and female vervet monkeys, and left them alone with a selection of toys, during which time they measured how often each vervet touched each toy.

The six toys were a police car and a ball ('male') a doll and a cooking pot ('female') and a book and a stuffed dog ('neutral'). Male vervets, it's claimed, played with boy toys and female vervets with girl toys, thereby proving that toy choice among human children must be innate and not the product of social conditioning.

Is this the most anthropomorphic experiment ever?

The baby doll is supposed to have elicited maternal instincts in the female vervet, but baby vervets do not look anything like dolls. In fact if there is one thing that a baby vervet looks like, with its tail, snout, four legs and fur is a toy stuffed dog. When it came to playing with the thing that looked most like a baby vervet – the toy dog – there is no difference between the male and female reaction.

In what sense can a vervet monkey be said to be playing with the toy police car at all? When children play with a police car, they imagine being a police officer, solving crimes, and chasing a criminal's getaway car. Do vervet monkeys even do the siren noise? In theory they could. Vervets have four distinct alarm calls for each of their four predators: eagles, leopards, pythons and baboons. Why not add one more for 'smash and grab at the jewellers'?

When a child plays with a toy pan or pot, he or she imagines that they're cooking. But we know that monkeys can't imagine cooking. If a monkey sees you cooking, and then has six million years to think about it, she still won't know why you are incinerating tasty treats, why you keep throwing good food in the fire.

Vervets play in the wild, of course, just not with toys. Because vervets haven't evolved to play with toys at all, and do not know what a toy is and are not much interested in a toy, they perform a quick, cursory inspection to check whether it is edible or useful in anyway, which, to the vervet, it is not. They pick it up, sniff it, put it down, and then go and do something else instead. What Alexander and Hines actually measured is how often each vervet touched the toys. This touching often involved no more than a tap or swipe of the toy. The total number of taps and swipes were added together, and it was found that males touched more toys more often than females touched them. Males usually claim first dibs on everything within a given territory. But we knew that already.

I had best declare an interest here. I submitted a rival paper to the science journal *Evolution & Human Behaviour* called 'Sex difference among chimpanzees in choice of fancy-dress costume', which was rejected in favour of Alexander and Hines's paper on toy choice in vervets.

I think my evidence base was just as strong. Given a random selection of fancy dress, the male chimps dressed as cowboys, the female chimps as squaws, the male chimps wore doctor's coats and female chimps tended to prefer nurse outfits. I also found evidence to suggest that chimpanzees pick up not only on human gender roles but also national stereotypes. The males dressed as Frenchmen with berets and onions round their necks, the females as Marie-Antoinette or Moulin Rouge can-can dancers.

I sent the peer-review panel at *Evolution & Human Behaviour* film footage of what happened when we set up a *Wizard of Oz* tableau, laid out fancy dress clothes on a yellow brick road and released three chimps into Oz. The female chimp dressed up as Dorothy, and the males as the Tin Man and the Scarecrow. I was pleased with the strength of the results. What counted against me with the peer-review panel was the discovery that I was charging an entrance fee for spectators. They were also suspicious that parts of the yellow brick road were heated and that the chimps were not so much dancing as hopping from foot to foot in agony.

Well, we can't all be gentlemen scientists in the nine-teenth-century tradition. We have to make science pay, and if by selling tickets to see an experiment being performed live I stand accused of having brought science to a wider public then I plead guilty as charged!

20. HOW MIND MAKES BRAIN

We are very different from our brains. If we tried to live on the brain's 100 per cent glucose diet we would quickly be in a diabetic coma, but the brain seems to thrive on it. The brain's version of efficiency is nothing like ours either. Presented with a photograph, the brain scatters the image to thirty different places the better to make sense of it as a single whole. But if I hand you a photo of my children, I hope you won't tear it into confetti and strew around the room before saying, 'Yes, I can see the likeness'.

In this chapter I want to focus on one other difference between ourselves and our brains, or rather a difference between the cranial biome the brain inhabits and the wide world that we inhabit.

Out in the world we distinguish sympathetic magic from science by the motto: correlation is not causation. We remind ourselves that just because two things happen together, it doesn't mean that one must have produced the other. But the nano-laws of the brain are different from the macro-laws of the world. For the brain, correlation *is* causation. Correlations of neuronal firing cause the map of synaptic connections to change. When two things happen simultaneously several times in a row they fuse together, forming a physical bond with each other. Neurons mark their shared history by shared topology. The brain does this not by way of sympathetic magic, but by the much more miraculous phenomenon of neuroplasticity.

In *The Organisation of Behaviour* (1949), Donald O. Hebb formulated what has since become known as Hebb's Rule, and which is usually expressed as:

> neurons that fire together, wire together,
> cells out of sync, don't link.

Hebb's Rule may have enjoyed more influence had it not been presented to the world in the form of hip-hop. This was bad for neuroscience and bad for rap, as demonstrated by Hebb's following lamentable use of that great lyrical no-no, the rhyming adverb:

> If the axon of one cell excites an adjacent cell repeatedly,
> Both will change their metabolism sympathetically.

Well, anyone could do that … indefinitely!

Mary Midgley gives the best translation of Hebb's hip-hop into prose, when she writes: 'minds change brains as much as brains change minds'. Habits of thought and feeling leave a physical trace, re-patterning neural networks, changing the brain's geography. Experience changes the layout of the brain, like rains shifting a river's sandbanks.

All this talk about how we are hard-wired turns out to be in need of a rethink. The notion that the brain clocks through pre-programmed modules like a washing machine on 'Cotton 40' will no longer do.

What I want to do now is to show that plasticity is not some special abracadabra of the brain, but only part of the wider developmental plasticity that made us who we are today. To do so I want to start by looking at the relation between skulls and brains.

Skulls and brains

We tend to think that intelligence is the glassblower and skull the molten glass. One deep breath and the skull swells like a pot-bellied vase, and you'd never guess we used to look like chimpanzees. But while nerve tissue can make small grooves in the skull such as the anterior condylar canal, it is simplistic to think that the shape of our skull is determined by the shape of the brain.

On BBC2's *Human Universe*, Brian Cox went to the Rift Valley where he lined up a three-million-year range of hominin skulls in a row. A row of skulls going from little to big. From a tiny million year old *Australopithecus* at one end, 'not much bigger than a chimpanzee brain', as you go through time the skulls get bigger and bigger. Half a million years ago there's a *Homo heidelbergensis* skull with quite a big brain. Then as you go through time again the skulls get bigger and bigger, and then from 200,000 years ago, here is an anatomically modern skull with 'a brain almost as big as mine'. Little skulls this end, big skulls that end. And that's evolution!

No, Brian, that's a xylophone.

The cerebellum, at the nape of the neck, has more neurons than the rest of the brain combined. If neuronal proliferation alone gave us our big skulls, our skulls would stick out at the back like a German helmet or a librarian's bun. The major cortical regions – frontal, temporal, parietal, occipital – are named after the adjacent bones of the skull. Not the other way around. (But that said, a fat and insistent nerve can score a small but significant groove in the skull, and the recent discovery one such groove – the condylar canal – in *Homo heidelbergensis* skulls has put Darwin's origin of language theory back in play, as we explored in the chapter 'Like Yesterday').

Of course, the way that a baby's cranial bones slide apart like tectonic plates to accommodate the rapidly growing brain makes it tempting to think that this is how evolution formed the skull. But this is to make the same blunder we saw Sigmund Freud make with his nutty theory of 'phylogenic recapitulation'. The blunder is to confuse ontogeny with phylogeny, to think the development of the individual follows the same pattern as the development of the species. The sliding bones of the baby's skull, incidentally, evolved for safe passage through the birth canal, not to accommodate exponentially proliferating brains .

Now I really like the Rift Valley evolutionary hypothesis *Human Universe* was sketching.[*] This concatenation of ecological events – deep lakes, evaporation, marooned populations and rapid speciation – seems a likely evolutionary scenario. But I'm

---

[*] Mark A. Maslin *et al.*, 'East African climate pulses and early human evolution', *Quaternary Science Reviews*, 2014.

not so keen on the xylophone. If bigger skull meant bigger brain, men would be cleverer than women, and tiny Isaac Newton would fetch up between *Homo ergaster* and *Homo heidelbergensis* on the skull xylophone. Yes, modern humans have bigger brains now than they did a million years ago, but this is partly because of the simple fact that our bodies are larger too. But what about the shape of these skulls? Don't they also become rounder and more human as they grow larger? Yes, they do but there are reasons for that which are more to do with looking cute than being smart. In the case of our brain-case, the shape of the skull may have been fashioned more by neoteny than neurons. Neoteny is the process by which our primate ancestors retained juvenile facial features as adults. If so, the change in our cranial capacity has less to do with being smart than it does with being baby-faced.

The *Oxford English Dictionary* defines neoteny as:

> The retention of juvenile characteristics in a (sexually) mature adult; esp. the appearance of ancestral juvenile characteristics in the adult stage of a descendant.

Humans and apes share a common primate ancestor. At some point about 3 to 4 million years ago, neotenous versions budded off the ancestral primate, creating *Homo*, the baby-faced ape. While Great Apes grow faces out to here – I mean literally *here*, exactly where you're holding the book – we keep the round skull of primate infancy, along with neotenous dentition and hairlessness. (Not me, alas. Only a box of Veet separates me from Cro-Magnon.)

But here is where form and function combine. Here's how the shape of the skull and the higher intelligence do indeed correlate, only it is not so simple as the glassblower and the vase. Hominin neoteny is not just about anatomy but about behaviour too.

Humans prolong development more than any other primates, and primates more than any other mammal. This allows more time for learning and play. We like to be able to look other animals in the eye, but we have to accept the fact that no other mammal was ever so cosseted as us. We are the Little Lord and Lady Fauntleroys of the animal kingdom. We spend forever in the nursery.

Our long dependency and painstaking acculturation freed us from a wide spectrum of selection pressures. This in turn allowed a greater variety of phenotypes to reach sexual maturity, which meant that when the winnowing came – Ice Age, flood, forest fire, disease – there was a broader spread to our bets.

But if neoteny was the making of us it was also our doom. It brought with it a curiosity and love of novelty to shame a jackdaw. Prolonged development allows for more exploratory play and investigation until you grow up with a taste for that sort of activity. We behave like juveniles (which may have something to do with the weary, slightly annoyed look that orangutans give us). Our love of novelty may explain why all species of *Homo* – neotenous like us – tend to be short-lived as species, if not as individuals. Curiosity killed the hominin.

The pessimistic mainstream take on why all other species of *Homo* are extinct is because they met with us. This is the premise, for example, of William Golding's unreadable novel *The Inheritors*. But a less pessimistic explanation might be that they shared our exploratory bent, our insatiable inquisitiveness, our Micawberism. Our headlong, often reckless improvisation may have been a characteristic shared widely across the genus, proving fatal to other hominin species such as *neanderthalensis, ergaster* and *heidel-bgergensis,* who died out bequeathing us fire, floral decoration, funeral rites, cooking, cross stitch and that formidable portfolio of bacterial genes which we never leave home without. For every successful catamaran expedition across the Indo-Pacific, twelve others sank at sea or were dashed on the rocks, with no survivors.

A cast of mind characterised by inquisitive, eager, probing, delving, exploratory behaviour is part of the inheritance that goes with the most prolonged juvenile development of any animals that ever lived. This shows that neoteny has to do both with the shape of our skull – round and large in relation to body size – and with the content of our heads as well. But not like the glassblower and the vase, but by a more winding path. It takes a whole host of evolutionary forces to give you a big head.

The social brain theory says that living in large complex groups got us thinking. Once we factor in neoteny to the social brain theory, however, we can be more specific and suggest that one particular aspect of social life above all others gave us our smarts. I think this aspect was that human life was organised around the care of the young. In this respect we more closely resemble African huntings dogs than chimpanzees. African hunting dogs bring back food that they regurgitate at the feet of nursing mothers and pups. In the den male and females dogs take turns to play, guard and feed the pups.[*]

---

[*] Wolfdietrich Kühme, 'Communal Food Distribution and Division of Labour in African Hunting Dogs', *Nature*, 1965.

The social life of African hunting dogs allows for newborn puppies as helpless as human babies. In humans, organising the life of the tribe around the care of the young amplifies the effects of neoteny.

* * *

The wild spotted hyena (*Crocuta crocuta*) provides a vivid example of how skulls change size and shape even when brain size stays constant. The spotted hyena's skull can radically change shape *within a single generation.*

In the wild, the skulls of adult hyenas are distinguished by a tall ridge on top, called a sagittal crest. The sagittal crest is the attachment site for extremely powerful jaw muscles. Attached to the top of the skull, like the cables of a suspension bridge, these jaw muscles exert maximum torque on the joints and tendons of large mammalian prey. Captive hyenas raised on a diet of diced meat, however, do not develop a sagittal crest because they do not develop jaw muscles strong enough to need anchoring. Never eating anything on the hoof, they retain the smooth rounded skull of a cub into adulthood. But let a round-headed captive hyena go feral, and her wild offspring will develop a sagittal crest as they hunt and grow. Having skipped a generation, the crest immediately reappears in the offspring as if it never went away!

This is an extreme example of the widespread phenomenon of *phenotypic plasticity.* Each organism inherits a constitution (its genotype) capable of a suite of alternative set-ups (its phenotype) in response to environmental cues. Phenotypic plasticity multiplies the role of the outside world in evolution, which is why its evolutionary role is downplayed, and why genes are preferred. Genes were once thought to be, in Dawkins' phrase 'sealed off from the outside world'. (Just as reality is now supposed to happen in 'the sealed auditorium of the cranium', according to Eagleman.)

A generation ago, it was taught that evolution could be summed as something like 'variation proposes, natural selection disposes'. Random genetic mutation threw up the variant, which took its chances in the lottery of life called natural selection. But that picture, satisfyingly neat though it was, has been falsified by evidence of organisms' flexible response to the world as they find it. Rather than let the local habitat pick them off for being unfit, organisms switch-hit to fit their habitat's fluctuating conditions. They deploy an astounding array of responses, up to and including growing a new-shaped skull. This switch-hitting and shape-shifting, this ringing the changes of the lifecycle is only sometimes called phenotypic or developmental plasticity. When the wind changes, it is called evolution.

* * *

I mentioned earlier that bigger bodies make bigger skulls. Sometimes called the 'meathead hypothesis', this accident of allometry is used to downplay the fact that the Neanderthal brain was bigger than our own. The only reason her brain was larger than ours, we tell ourselves, is because her body was bigger. But as a proportion of her body mass, so the argument goes, her Neanderthal brain was smaller. In this way we feel safe in dismissing claims of Neanderthal intelligence. But we can't write off Cro-Magnon in the same way. Compared to modern humans, Cro-Magnon had more brains, less brawn.

The earliest Cro-Magnon people, with brains bigger than our own, produced stunning paintings in their caves, but did not write symphonies nor build computers. All that we have accomplished since then is the product of cultural evolution based on a brain of unvarying capacity.*

Unvarying capacity. This is the big mystery: long before humans made the Great Leap Forward, *everything was already in place, we already had the big brain.* We were like the Scarecrow in *Wizard of Oz* who sings 'If I Only Had A Brain' never realising he has a fine one all along.

---

* Stephen Jay Gould, *The Panda's Thumb* (Chapter 4), 1990.

21. THE ORIGIN OF MIND

Around 700 million years ago, we find the first ever sensori-motor systems laying down the first ever memories in the first ever minds. The memories belong to marine invertebrates on the ocean floor. What they remember is the chemical ambience of the water where they were a few seconds ago. They compare this memory with where they are now, before deciding whether to swim straight ahead or to do a tumble-turn. They navigate by chemical concentration gradients, seeking or avoiding chemicals, in a type of movement called chemotaxis.

In 'Animal Evolution and the Origins of Experience', Peter Godfrey-Smith goes back to the Cambrian Explosion fossil record to show how 'mind evolved in response to other minds', and 'bodies evolved in response to other minds'.

For most of the Ediacaran Epoch (635–541 million years ago) nervous systems don't put in much of a shift. After a light spot of metabolic homeostasis, the mycoplasma unwind with a lazy backstroke. Even at this stage, however, these multicellular organisms establish a rule that will hold will hold for the rest of time: right from the start there is no brain without brawn.

As Peter Godfrey-Smith says 'only organisms with neurons also have muscle cells'.*

The Ediacaran Eden is the peaceful flotation tank of evolution. Life on earth will never again be so Zen. It's a steam bath, a jacuzzi. Ediacaran marine invertebrates are benthic grazers on microbes, filter feeding on whatever comes their way. Nobody eats anybody else. There's no predator–prey action. No fossils of half-eaten organisms come from the Ediacaran.

But all that changes with the Cambrian Explosion, which inaugurates both the Cambrian Period (541–485 million years ago) and the three-hundred-million year long Paleozoic Era. The Cambrian Explosion does for the Ediacaran what Prohibition does for Chicago. Now, the wild rumpus begins, and there goes the neighbourhood! Signs of predation are everywhere. The Cambrian fossil record shows a body-count going up and up. Everyone is packing some kind of piece: a claw, a spike, armour, pincers and shivs. With predators lurking in the murky depths, you need to react in real time to what is going on around you.

'New kinds of bodies appear', says Peter Godfrey-Smith. 'For the first time in history the details of what were going on around start to matter.' Now the first antennae go up. Now come the first getaway fins and spines. Now come the first real-time sensory-motor complex arcs. Now come what Peter Godfrey-Smith characterises as 'image-forming eyes for tracking other animals at a distance'.

To survive the Cambrian you need a body capable of

---

* Peter Godfrey-Smith, 'Mind, Matter and Metabolism', *Journal of Philosophy*, 2017.

anticipating or responding to whatever dark and devious proto-thoughts are going on in the *Haikouichthys* swimming towards you. This is what he means by saying that both mind and body evolve in response to other minds.

* * *

In *The Sea Around Us* (1951) Rachel Carson, the famous ecologist and marine biologist, describes how the composition of our blood faithfully records the precise ratios of sodium, potassium and calcium in Cambrian oceans. Since the brain works by means of pumping sodium and potassium ions, it is tempting to speculate that the ability of the mind to bring just one thing at a time into consciousness copies chemical patterns used in taming the pre-Cambrian benthic blitz. In 'Mind, Matter and Metabolism' Peter Godfrey-Smith describes how life originated in a 'molecular storm':

> [T]he origin of life went not from simple to complex but from disorderly to orderly ... The evolution of life was a matter of channeling and taming a sea of interactions not taking a few simple interactions and stringing them together.

Perhaps organisms mimicked or accidentally ingested the ambient chemical processes. Inside the cellular membrane this biochemistry – as if conscripted – carried on doing what it had been doing before, but this time it found itself taming not an ocean but incoming sensory overload into one coherent and orderly mental workbench.

From a blizzard of mental activity the mind brings just one or two things into consciousness, and lays them on the mental workbench or 'global workspace'. Within this noisy chemical bombardment the brain finds its scallop-shell of quiet. We say 'I can't think' when there is too much going on in our heads, not when there's too little. The exception to this rule is Homer Simpson's sad lament: 'I want to be left alone with my thought'. Is it possible that flow patterns in the brain mimic biochemical patterns from ancestral seas?

This is not a hypothesis that Peter Godfrey-Smith himself entertains, I hasten to add. In fact, in 'Cephalopods and the Evolution of the Mind',[*] he suggests a more convincing hypothesis than mine for how the mental workbench evolved. It may have come about he suggests by the demands of extractive foraging, and by 'the way the need to perform novel combinations of operations presses events into consciousness ... [because] they must maintain information for explicit use over a period of time.' In his 'Cephalopods' paper, Peter Godfrey-Smith is intrigued by the way that octopuses achieve a single integrated consciousness despite the fact that two-thirds of their neurons are in their arms:

> Although many processes are going on in parallel in our brains, what we are aware of at any time in a single integrated scene ... The familiarity of the fact that conscious experience is integrated in this way disguises the fact that this integration is a significant achievement for our brains.

---

[*] Peter Godfrey-Smith, 'Cephalopods and the Evolution of the Mind' *Pacific Conservation Biology*, 2013.

If it is a significant achievement for us, then how much more so for the decentralised octopus! The last common ancestor of human and octopus shimmied along the Ediacaran ocean floor, about 600 million years ago. Since they have a completely different evolutionary trajectory to us, this makes them, he says

> an *independent experiment* in the evolution of large and complex nervous systems ... Where their minds differ from ours, they show us another way of being a sentient organism.

Because octopuses live short lives, only a few years at best, neoteny has nothing to do with the evolution of their smarts as it has for ours. And since octopuses are not very sociable, you can't ascribe their intelligence to the social brain theory, which is another theory for the origin of human intelligence. And yet the octopus demonstrates sophisticated levels of planning and tool use in excess of even the higher primates.

About twenty years ago, for example, marine biologists in the Indonesian Sea first noticed octopuses carrying pairs of coconut shells, like stacking bowls. No-one knew where they were taking them or what they were going to do with them when they got there. It remained a mystery for a couple of decades, but a recent paper appears to have solved the enigma. The octopuses, it turns out, are using the coconut shells as a portable storm shelter. When cyclones roll rocks and stones along the ocean floor, the octopuses climb into the coconut halves, and seal them shut with their strong arms. It's a self-assembly spherical storm refuge.

I can't tell you how depressed I was by this finding. For as long as I live, a part of me will always hope that the real reason octopuses carry pairs of coconut shells is for the sarcastic ridicule of seahorses.

22. ATTACK OF THE KILLER SCI-FI

Thirty miles from Paris, a length of elevated ferroconcrete in the Essonne sedgeland is all the space the future had for Aérotrain France. In Seville, a derelict monorail stops in mid-air. In a Goan junkyard, vines, weeds and creepers curl around the grimy passenger pods of Sky Bus Metro. A timber shed in Milngavie, Scotland marks the last known whereabouts of the Bennie Railplane ('Swift. Safe. Sure.'). These abandoned monorails are monuments to old-time futurism or retro sci-fi.

The route of the abandoned Aérotrain France monorail is now a wildlife conservation area, home to the yellow spotted black tiger moth and the little bittern. The elevated ferroconcrete monorail track in the middle of these wetlands gives you a feeling similar to when you stumble upon Aztec or Mayan ruins in the forest. Here is a structure designed by people with a completely different idea of the world to us. It is a sort of ghostly apparition. And indeed, mingled among the notes of the bittern booming in the sedge, you can also hear the ghosts of monorail enthusiasts denouncing doubters as Luddites.

Today's Luddites are those of us who are sceptical about staples of retro-sci-fi such as strong AI, uploading consciousness, brain-decoders, dream-decoders, polygraphs, wisdom implants and the Singularity – all of which are expressions of belief not in science but in a kind of magical scientism

The tension within neuroscience between visions of the future based on science and ones based on magical scientism is part of a wider social struggle to escape the attack of the killer retro sci-fi. By retro sci-fi, I mean visions of the future that were produced in the Age of Cheap Oil and before Rachel Carson's *Silent Spring*, a handy place marker for the start of the environmental movement. Retro sci-fi stops anything new being done. Many different parts of society struggle to escape its stranglehold, from conservation and agriculture, to town planning and architecture.

I have a fantasy in which I go back in time forty years and tell a modernist or postmodernist architect, Le Corbusier or Richard Rogers, say, that what really dates his buildings in twenty-first century eyes is their lack of access ramps and bike racks. He looks at me as if I'd said that the problem is the lack of arrow slits or anywhere to store a siege catapult.

Le Corbusier or Rogers retort that if the twenty-first century has rejected his bold, forward-thinking cityscapes, then it must have suffered a failure of nerve, vision, ambition, imagination, a failure that left it unable to dream the impossible dream or to appreciate the production values on Kraftwerk's *Autobahn*. In other words, I imagine Le Corbusier to respond much like Norma Desmond in *Sunset Boulevard*, when told that she used to be big. 'I am big', she replies. 'It's the pictures that got small.'

It's our shortcomings that are to blame for abandoning the architectural future-scapes. We gave up on thinking big. But that is a long way from the truth. The real reason their grandiose future-scapes don't fit the way we live now is because we came up with a more sophisticated, more human, more inclusive society, which was completely beyond their ability to imagine, and which the bike rack and the access ramp symbolise rather well. The bike rack symbolises our change in thinking about the urban environment, in particular our doubts about whether the best use of streets is efficient through-put of maximum tonnage of traffic volume. The bike rack also represents an appreciation of scientific evidence about the link between fossil fuels, atmospheric carbon, and global warming. The access ramp commemorates achievements made by disabled campaigners for equal human rights, which were set in stone in the great Disability Discrimination Act of 1995. The access ramp represents an unprecedented advance in thinking about difference and inclusion.

I'm sure the Colossus of Rhodes was a wonder to behold, but the access ramp is a far greater testament to the human spirit. (I'd say the same for the Statue of Zeus at Olympia, but the for the fact that some classicists are of the opinion that his left hand held a scroll which bore the inscription 'Buggies Welcome Here'.)

Just as it would be fun to put postmodern architects' noses out of joint, what time traveller would turn down a chance to tell monorail designers that the only amazingly fast thing about their futuristic transportation system was the speed with which it ended up in the scrapyard?

Space colonies are an excellent illustration of how retro-sci-fi celebrates the victory of magical scientism over science. The belief that you can by any self-sustaining process live on Mars is called scientific, whereas those of us who cling to the outmoded belief that the entropy of a closed system increases with time are Luddite future-phobes. Things have come to a pretty pass when anyone who repudiates the Second Law of Thermodynamics is pro-science, whereas those who accept the Second Law are anti-science. But there is no better illustration of the fork in the road we have come to, the split between visions of the future based on retro sci-fi and ones based on science.

Mars is entropic because it has no soil, only regolith. Soil is composed of organic matter and the red planet is dead. I'll watch just about anything that stars Matt Damon, but his film *The Martian* has a contemptuous attitude to soil. The moral of the film, unless I've missed something: is gravel + faeces = soil. Just shit on some regolith and hey presto – soil! May I point out that the only reason he can grow anything in the cadmium dust of Mars is 100 per cent due to life on earth. 'If you follow the energy', as environmental author Richard Manning says, 'you will eventually end up in a field somewhere." And that field will somewhere on earth. The usefulness of astronaut plop as fertiliser is entirely traceable to Iowan wheat fields and Sinaloan peach orchards.

News reports claiming that Dutch scientists have replicated Matt Damon's 'plopiculture', which means that we will be able to farm on Mars, omitted to mention that these Dutch experiments were funded by an organisation called Mars One, which hopes to set up space colonies. And of course the crops were

---

* Richard Manning, 'The Oil We Eat', *Harper's*, 2004.

not grown in Martian dust at all, but in a soil engineered by NASA using organic material from the only place in the known universe where you can get your hands on that type of thing, Planet Earth.

The Darién Gap versus the Pan-American Highway

Seattle is one of the few cities still with an extant straddle-beam monorail, and it's the place to go if you want to understand why this sci-fi transportation system is a thing of the past. You don't need to buy a ticket, because you won't be boarding the monorail itself. Simply walk its route from, say, the Westlake Centre to Fifth Avenue. You will be walking in shadowy cold and breathing a gritty, gusty air. The monorail's ferroconcrete infrastructure puts a lid on the Seattle's magnificent views and fresh Pacific breeze. It casts a heavy shadow over the shops, streets, flats, parks, bars and cafes in its path. When monorails were dreamt up, the pedestrian lifestyle was thought to be on the way out. Soon we would all be flying in personal rocket-ships to gigs, where bands play innovative synth-based music made possible by the Fairlight Computer Musical Instrument, or Fairlight CMI for short. Cities as places for living were over. We would travel to and from cities but we the idea that we would actually live in them was as quaint as gas-lights. No-one would live in cities except rats, janitors and super villains. Suburbs were the future. Cheap oil created suburbs, and many of our visions of the future, such as intensive agriculture. Farming output would be produced in ever-larger economies of scale, would depend on ever-greater quantities of oil-based insecticides and pesticides, and ever-increasing food miles from field to fork. Unlike coal, it seemed there was no harm in burning petroleum, diesel and no harm in oil-based agricultural inputs and no harm in instant food.

The food pill was a staple of 1970s sci-fi films. No need to bother with preparing food. No more cooking. One pill and that was you done for the day. Food couldn't be faster than that. This retro sci-fi fantasy makes a comeback in *Homo Deus*, where we learn that in the future there will be 89 per cent fewer bakers than there are today.

Once again, the trend is completely in the other direction, just as we saw with security guards. The last decade, according to the Office of National Statistics, has seen a year-on-year increase in the number of new bakeries opening up. Ah, but aren't these automated bakeries staffed by bread-making robots? Not really, because craft bakeries are the growth area. The Craft Bakers Association calculates that 40 per cent of the people employed in craft baking are bakers. The total number of craft bakers is set to increase to match volume growth in bread, cakes and pastries. Who'd have thought the future would be less rocket ships and more braided sourdough? That instead of food pills, there would be seeded spelt, a slow-food movement and an enormous waiting list for allotments.

* * *

Our heads are battlegrounds between past and future. We find ourselves at a moment of struggle between rival visions of the future, ones suggested by the Age of Cheap Oil's retro sci-fi, and ones based on science. This is played out in many places and many ways. It's the struggle between fast food and slow food, between ten-lane urban freeway and Home Zone, Cargill and the Soil Association, helipad and bike rack, monorail and access ramp.

If we don't break old-time futurism's stranglehold on the future, then the future will be like 1970s sci-fi: the sets will wobble, the genders will be stereotyped and there will be a thin-ly-disguised fear of Russian expansionism behind every story.

Just as the Darién Gap's impenetrable jungle blocks the route of the Pan-American Highway, so environmentalism impedes retro-sci-fi's route to the future. But environmental victories are being rolled back, and global warming could dry out the Darién swamps enough to allow the highway finally to be bulldozed through after all and the approach roads to Neuropolis begun.

23. THE NEUROBABBLE AND THE DAMAGE DONE

Neurobabble damages us more than we know by making us distrust ourselves, and each other. This has a chilling effect on democratic participation. If our judgement cannot be trusted, if our consciousness is false, if we are anti-social killer apes whose homicidal instincts are barely restrained by a slender cortical add-on, then who are we to decide anything?

And what is the accumulated psychological effect of all this? No doubt our global depression epidemic has many causes, but it can't help being told that your body is some sort of useless appendage, that your sense of self, 'you', your joys and sorrows are all an illusion and that you can be uploaded onto a thumb stick.

There are enough forces working to dehumanise us already, so why rush into the embrace of the neural dehumanisers? Why do we give our humanity away so easy? Is it because we are told it's the scientifically correct thing to do? If so, that is desperately sad because the vast and belittling claims advanced by neurobabble are bad biology and will be until the tree frog appears on the brain scanner's screen.

We are always told the excellent thing about scientific method is that it controls for bias. But it controls for some bias better than others. The ready uptake of ideas about uploading consciousness, say, or the acceptance of the fantastic notion that smiling evolved from snarling, have to do with the fact that the scientific imagination was primed for such a find, because such a find would confirm a lot of political and philosophical assumptions, the very assumptions which sent them to look for those things in that place and in that way. As Ray Tallis says: '[neuroscience] treats individuals as passive respondents to stimuli and then discovers that they are passive respondents to stimuli.'

This crude circularity elevates contempt to a scientific technique. If you reduce people enough then it becomes much easier to design quantifiable, measurable experiments. The only problem is that by the time you zoom the picture into such big-dot resolution, the contempt extends to science too, and what you are left with is a sort of positivist pseudo-science, which tends not to be able to explain very much.

A defining characteristic of the current crop of science popularisers, for example, is to swap explanatory complexity for scale. That is why they bang on about immense expanses of time and space. Never mind the quality feel the width. No astronomy program is complete without the roll-call of our insignificance, the tedious litany of a trillion trillion planets orbiting a trillion trillion stars. The vasty vastness of deep geological time is supposed to silence doubts about the explanatory thinness of the reptilian brain hypothesis. But scale is no substitute for exploration of the complex interactions of human behaviour.

A scientific consensus is emerging that we have wildly underestimated the degree to which the neural structures of the mind are environmentally and socially produced. This has major social implications. The world we build makes us. We've always known this of course, but there is now a scientific basis for saying that the damage done to the ecosystem is damage done to the soul.

Fantasies of evolving into 'non-biological beings' couldn't come at a worse time. Now more than ever we need to understand how, deeply intertwined with the biosphere, there is no remedy for ecological collapse without this understanding.

When seawater gets above a certain temperature, coral expel their zooxanthellae, the photosynthetic algae that give the reef its colour. But if the world is colourless anyway, then what does it matter if the Great Barrier Reef's coral is now polystyrene white for seven hundred miles? Aren't we about to shuck off nature and become non-biological beings, anyway?

The way we look at the brain is the blowback of a particularly stunted and atomised way of looking at the world. The conceptual razing of the outside world in neuroscientific dogma is an offshoot of the physical razing of the living world by way of clear cutting, scallop dredging, soil skimming and coral bleaching. Neuroscientific dogma also echoes a widespread attack on people's sense of agency and identity, both in the workplace and in public life in general.

The moral of this story is that brain-imaging is not the step-change we are being sold, not least because fMRI does not show our brains in action, or anything like it. Now is *not* one of those times, like when Galileo turned his telescope on the heavens and we swapped astrology for astronomy. Now is *not* a moment, like when van Leuwenhoek's microscope revealed the secret world of bacteria. Now is in fact the very opposite of one of those times. Now is the last hurrah of an old philosophy that sees us as not really of the earth, as not belonging here, as being too good for this place. The last hurrah of the machine metaphor. The last hurrah of old-time sci-fi. The very moment when we think we see cold, hard reality, is the very moment that we are in fact at our most cloudy and mystical, least clear and scientific, while the world slips silently from sight, losing definition, colour, existence.

We face our greatest challenge in the long emergency of global warming, resource depletion and mass species extinction. The decisions we make in the coming years have a better chance of being good decisions if we have a clear understanding of who we are, who we are not, and of what we yet might be.

BIBLIOGRAPHY

Barrett, Louise, *Beyond The Brain:
How Body and Environment Shape
Animal and Human Minds*, Princeton
University Press, Princeton, NJ, 2011.

Beck, U. *et al.*, 'The invention of trousers
and its likely affiliation with horseback
riding and mobility: A case study
of late 2nd millennium BC finds
from Turfan in eastern Central Asia',
Quaternary International, Vol. 348, pp.
224–35, 2014.

Berkley, George, *A Treatise Concerning The
Principles of Human Knowledge*, Dublin,
1710.

Berlin, Isaiah, *The Age of Enlightenment*,
Meridian, New York, 1984.

Bisiach, E. & Luzzatti, C., 'Unilateral
Neglect of Representational Space',
Cortex, vol. 14, no. 1, pp. 129–33,
March 1978.

Borota, D., Murray, E. *et al.*, 'Post-study
caffeine administration enhances
memory consolidation in humans',
Nature Neuroscience, vol. 17, no. 2, pp.
201–3, Feburary 2014.

Boswell, James, *The Life of Johnson*, Penguin
Classics, London, 2008.

Castellano, Claudio, 'Effects of
caffeine on discrimination learning,
consolidation and memory in mice.',
Psychopharmacology, Vol. 48, No. 3, pp.
255–60, January 1976.

Coleman, Terry, *The Railway Navvies*,
Penguin, London, 1968.

Collins, Billy, *Taking Off Emily Dickinson's
Clothes*, Picador, London, 2000.

Crick, Francis, *The Astonishing Hypothesis*,
Simon & Schuster, 1995.

Darwin, Charles, *The Expression of
the Emotions In Man and Animals*,
reprinted by Oxford University Press,
Oxford, 1998.

———, 'A Biographical Sketch of an
Infant, *Mind: a Quarterly Review of
Psychology and Philosophy*, vol. 2, no. 7,
pp. 285–94, July 1877.

Dewey, John, *Experience and Nature*, Dover
Publications, New York, 1998.

———, 'The Reflex Arc Concept in
Psychology', *Psychological Review*, vol.
3, pp. 357–70, 1896.

Dickens, Charles, *David Copperfield*,
Wordsworth Editions, Ware,
Hertfordshire, 1992.

Doidge, Norman, *The Brain that Changes
Itself*, Penguin, London, 2007.

Domes, G., Heinrichs, M., Michel, A., Berger, C. & Herpertz, S. C., 'Oxytocin improves "mind-reading" in humans,' *Biological Psychiatry*, vol. 61, no. 6, pp. 731–3, March 2007.

Donald, Merlin, *A Mind So Rare: The Evolution of Human Consciousness*, W. W. Norton, New York, 2001.

Eagleman, David, *The Brain: The Story of You*, Canongate Books, Edinburgh, 2015.

Eklund, A., Nichols, T. E. & Knutsson, H., 'Cluster failure: Why fMRI inferences for spatial extent have inflated false-positive rates', *PNAS*, vol. 113, no. 28, pp. 7900–05, July 2016.

Finger, Stanley, *Origins of Neuroscience: A History of Explorations into Brain Function*, Oxford University Press, Oxford, 1994.

Freeman, Walter J, 'The Physiology of Perception', *Scientific American*, February 1991.

Freud, Sigmund, *Totem & Taboo: Some Points of Agreement Between the Mental Life of Savages and Neurotics* (trans. James Strachey), Routledge & Paul, London, 1950 [1913].

Gauthier, Tarrand, 'FFA: a flexible fusiform area for subordinate-level visual processing automatized by expertise', *Nature Neuroscience*, Vol. 3, no. 8, pp. 764–9, August 2000.

Gibson, James J., *The Ecological Approach to Visual Perception* (New Edn), Routledge, London, 1986.

Gombrich, E. H., *Art and Illusion*, Phaidon Press, London, 2002.

Godfrey-Smith, Peter, 'Animal Evolution and the Origins of Experience', in Livingstone Smith, David (ed.), *How Biology Shapes Philosophy: New Foundations for Naturalism*, Cambridge University Press, Cambridge, 2017.

————, 'Cephalopods and the Evolution of the Mind', *Pacific Conservation Biology*, vol. 19, no. 1, pp. 4–9, January 2013.

————, 'Mind, Matter and Metabolism', *Journal of Philosophy*, (In press) 2017.

Gould, Stephen Jay, *Ontogeny and Phylogeny*, Harvard University Press, Cambridge, MA, 1977.

————, *Rocks of Ages: Science and Religion in the Fullness of Life*, Ballantine, New York, 1999.

————, *The Panda's Thumb: More Reflections in Natural History*, Penguin Books, 1990 .

Gregory, R. L. & Gombrich, E. H. (eds), *Illusion in Nature and Art*, Duckworth, London, 1973.

Harari, Yuval Noah, *Homo Deus: A Brief History of Tomorrow*, Harvill Secker, London, 2016.

Hazlitt, William, *Table Talk: Essays on Men and Manners*, Oxford University Press, London/New York, 1902.

Hebb, D. O., *The Organisation of Behaviour: A Neuropsychological Theory*, Wiley, New York, 1949.

Hunt, S., Bennett, A. T. D, Cuthill, I. C. & Griffiths, R., 'Blue tits are ultraviolet tits', *Proc. Biol. Soc.*, vol. 265, no. 1395, pp. 451–5, March 1998.

Jablonka, Eva, & Lamb, Marion J., *Evolution in Four Dimensions: Genetic, Epigenetic, Behavioral, and Symbolic Variation in the History of Life* (rev. edn), MIT Press, Cambridge, MA, 2014.

Jemison, Mae, *Find Where The Wind Goes: Moments From My Life*, Scholastic Press, New York, 2001.

Jespersen, Otto, *Language: Its Nature, Development, and Origin*, George Allen & Unwin, London, 1968.

Jung, C.-R., Lin, Y.-T. & Hwang, B.-F., 'Ozone, particulate matter, and newly diagnosed Alzheimer's disease: a population-based cohort study in Taiwan', *Journal of Alzheimer's Disease*, vol. 44, no. 2, pp. 573–84, 2015.

Kosfeld, M., Heinrichs, M., Zak, P. J., Fischbacher, U. & Fehr, E., 'Oxytocin increases trust in humans', *Nature*, 435, pp. 673–6, June 2005.

Kühme, Wolfdietrich, 'Communal Food Distribution and Division of Labour in African Hunting Dogs', *Nature*, vol. 30, no. 205, pp. 443–4, Jan. 1965.

Lange, Carl & James, William, *The Emotions*, Williams & Wilkins, Baltimore, MD, 1922.

Lee, K.-J., Inoue, M., Otani *et al.*, 'Coffee consumption and risk of colorectal cancer in a population-based prospective cohort of Japanese men and women', *International Journal of Cancer*, vol. 121, no. 6, pp. 1312–18, Sept. 2007.

McGilchrist, Iain, *The Master and His Emissary: The Divided Brain and the Making of the Western World*, Yale University Press, New Haven and London, 2009.

Macmillan, Malcolm, *An Odd Kind Of Fame: Stories of Phineas Gage*, MIT Press, Cambridge, MA, 2000.

Maher, Barbara *et al.*, 'Magnetite pollution particles in the human brain', *PNAS*, vol. 113, no. 39, pp. 10797–801, 2016.

Manning, Richard, 'The Oil We Eat: Following the food chain back to Iraq', *Harper's*, February 2004.

Marcus, Gary, *Kluge: The Haphazard Construction of the Human Mind*, Faber & Faber, London, 2008.

Maslin, Mark A. *et al.*, 'East African climate pulses and early human evolution', *Quaternary Science Reviews*, 101, pp. 1–17, Oct. 2014.

Meeks, Thomas W. & Jeste, Dilip V., 'Neurobiology of Wisdom' *Archive of General Psychiatry*, vol. 66, no. 4, pp. 355–65, (2009).

Midgley, Mary, *Beast and Man: The Roots of Human Nature*, Routledge, London/ New York, 2002.

————, *Are You An Illusion?*, Routledge, London, 2015.

Morgan, C. Lloyd, *Introduction to Comparative Psychology*, W. Scott Ltd, London 1894.

Newman, Robert, *The Entirely Accurate Encyclopaedia of Evolution*, Freight Books, Glasgow, 2015.

O'Keefe, J. & Dostrovsky, J., 'The hippocampus as a spatial map. Preliminary evidence from unit activity in the freely-moving rat', *Brain Research*, vol. 34, no. 1, pp. 171–5, November 1971.

Pinker, Steven, *How The Mind Works*, Norton, New York, 1997.

Popper, Karl, *Conjectures and Refutations: The Growth of Scientific Knowledge*, Basic Books, New York, 1962.

Puglia, Meghan *et al.*, 'Epigenetic modification of the oxytocin receptor gene influences the perception of anger and fear in the human brain', *PNAS*, vol. 112, no. 11, pp. 3308–13, 2015.

Putnam, Hilary, *The Collapse of the Fact/ Value Dichotomy and Other Essays*, Harvard University Press, Cambridge, MA, 2002.

Ramachandran, VS, & Blakeslee, Sandra, *Phantoms in the Brain: Probing the Mysteries of the Human Mind*, William Morrow, New York, 1998.

Renner, M. J. & Rosenzweig, M. R., *Enriched and Impoverished Environments: Effects on Brain and Behaviour*, Springer-Verlag, New York, 1987.

Robinson, Marilynne, *Absence of Mind: The Dispelling of Inwardness from the Modern Myth of the Self*, Yale University Press, New Haven, CT, 2010.

——, *The Death of Adam: Essays on Modern Thought*, Houghton Mifflin, Boston, 1998.

Russell, Bertrand, *A History of Western Philosophy*, Simon and Schuster, New York, 1945.

——, *Mysticism and Logic*, W. W. Norton, New York, 1929.

Russell, Stuart 'Take a stand on AI weapons', *Nature*, Vol. 521, No. 7553, May 2015.

Satel, Sally & Lilienfeld, Scott O., *Brainwashed: The Seductive Appeal of Mindless Neuroscience*, Basic Books, New York, 2013.

Schultz, Duane P. & Schultz, Sydney Ellen, *A History of Modern Psychology*, Thomson/Wadsworth, Belmont, CA, 2012.

Shamay-Tsoory, S. G. *et al.*, 'Intranasal administration of oxytocin increases envy and schadenfreude (gloating)' *Biological Psychiatry*, vol. 66, no. 9, pp. 864–70, November 2009.

Stringer, Christopher & Gamble, Clive, *In Search of the Neanderthals: Solving the Puzzle of Human Origins*, Thames and Hudson, New York, 1993.

Swaab, Dick, *We Are Our Brains: From the Womb to Alzheimer's*, Allen Lane, London, 2014.

Tallis, Raymond, *Aping Mankind: Neuromania, Darwinitis and the Misrepresentation of Humanity*, Routledge, London, 2014.

Thelen, Esther & Smith, Linda B., *A Dynamic Systems Approach to the Development of Cognition and Action*, MIT Press, Cambridge, MA, 1994.

Todes, Daniel P., *Ivan Pavlov: A Russian Life in Science*, Oxford University Press, Oxford, 2014.

Tolman, Edward C., 'Cognitive Maps in Rats and Men', *Psycvhological Review*, vol. 55, no. 4, pp. 189–208, July 1948.

Vul, Ed *et al.*, 'Puzzlingly High Correlations in fMRI Studies of Emotion, Personality, and Social Cognition', *Perspectives on Psychological Science*, vol. 4, no. 3, pp. 274–90, May 2009.

Vygotsky, Lev, *Mind In Society: The Development of Higher Psychological Processes*, Harvard University Press, London/Cambridge, MA, 1978.

Wall, Patrick, *Pain: The Science of Suffering*, Weidenfeld & Nicolson, London, 1999.

Wells, Andrew, *Rethinking Cognitive Computation: Turing and the Science of the Mind*, Palgrave Macmillan, Basingstoke, 2006.

Wertheim, Margaret, *Pythagoras's Trousers: God, Physics, and the Gender War*, W. W. Norton, New York, 1997.

Wismer, Sharon *et al.*, 'Variation in Cleaner Wrasse Cooperation and Cognition: Influence of the Developmental Environment?', *Ethology*, vol. 120, no. 6, pp. 519–31, March 2014.

Zimmer, Carl, *Soul Made Flesh: The Discovery of the Brain and How it Changed the World*, Heinemann, London, 2004.

INDEX

THANKS AND ACKNOWLEDGEMENTS

Myles Archibald, Clare Alexander, Jonathan Harvey, Claire Price, Bill McCabe, Nicky Fijalkowska, Tom Cabot, Ed Smith and Vesselina Newman.